THE HIGH-PERFORMANCE MIND

MASTERING BRAINWAVES FOR INSIGHT, HEALING, AND CREATIVITY

ANNA WISE

Jeremy P. Tarcher/Penguin
a member of
Penguin Group (USA) Inc.
New York

Most Tarcher/Penguin books are available at special quantity discounts for bulk purchase for sales promotions, premiums, fund-raising, and educational needs. Special books or book excerpts also can be created to fit specific needs. For details, write Penguin Group (USA) Inc. Special Markets, 375 Hudson Street New York, NY 10014

Jeremy P. Tarcher/Penguin
a member of
Penguin Group (USA) Inc.
375 Hudson Street
New York, NY 10014
www.penguin.com

Published simultaneously in Canada

LIBRARY OF CONGRESS CATALOGING-IN-PUBLICATION DATA
Wise, Anna.
The high-performance mind : mastering brainwaves for insight,
healing, and creativity / Anna Wise.
p.   cm.
Includes index.
ISBN 0-87477-850-6
1. Biofeedback training.   2. Alpha rhythm.   3. Behavior
modification.   4. Success—Psychological aspects.   I. Title.
BF319.5.B5W57   1996   95-9249 CIP
153—dc20

Design by Fritz Metsch
Illustrations by Barbara Walsh
Revised illustrations for paperback edition by Terry M. Hardy
Printed in the United States of America
15 17 19 20 18 16 14

This book is dedicated with enduring love and gratitude to my parents and especially to my son, John Michael Wise. Thank you, John, with all my heart for your love, support, help, and understanding. Your belief in me has often kept me going. You've celebrated the light and shared the darkness. I love you "to infinity."

ACKNOWLEDGMENTS

I would like to acknowledge and honor the late C. Maxwell Cade, without whom this work would not have been possible; his wife and fellow teacher, Isabel Cade; the late Geoffrey Blundell, whose technical genius was invaluable; and Neil Hancock for his software expertise. I would also like to give appreciation and acknowledgment to Elizabeth St. John for being with me on this journey and for more than thirty years of friendship, healing, and support; and to Nancie Lin, my Chinese sister, wo hen gan ji ni. Xiexie.

My deepest appreciation and gratitude also go to my healers and spiritual teachers, for keeping me together—body and soul—especially Master Wu, Swami Satyananda Saraswati, Mary Simpkins, Masters Donald and Cheryl Lynne Rubbo, Yat Ki Lai, Dr. Carole Warnes, Dr. Michael Osborn, Dr. Howard Kornfeld, Peter Fabian, Gale Ulvang, and Reverend Cecil Williams. A profound and grateful acknowledgment goes to the Matsu Temple in Lukang, Taiwan. Matsu, thank you.

Deep appreciation also to those essential people who have helped in so many diverse and crucial ways—Bob Gordon, Mark Carson, Nancy Lunney-Wheeler, Rob Kall, Tom and Terri Collura, and Joel Fotinos and especially Jeremy Tarcher, a true visionary. Thank you for your faith. And finally, I would like to recognize and thank all of my students and clients over the years, who have given me the opportunity to practice doing what I love. You are the life's breath of my work, and without you, it would not have been possible.

CONTENTS

———————— ■ ————————

At some point, all of us have sat at our computers overwhelmed by the workload, not knowing where to start while our minds are filled with so much chatter or worry that it can cause trouble relaxing or going to sleep at night. Some of us begin to feel that we have lost our creative edge or perhaps *know* there is something creative deep inside trying to get out, but we can't quite get in touch with it! Or maybe you crave being more in touch with your "higher self" or with God but don't know how, and perhaps you have been feeling as though your life has lost its purpose.

Whether they seem enormously overwhelming or just irritating, you can discover how to address all of these common problems through *The High-Performance Mind,* where you will learn how to apply your brainwaves to find the solutions to these and other situations that plague our modern high-stress society. You will also learn the science and the art of meditation and how to develop the mastery needed to meditate whenever and for whatever purpose required, whether it is for self-healing, stress management, creativity, peak performance, personal growth, or spirituality. If you already know how to meditate, this book will help you fine-tune your practice and take you even deeper inside yourself.

The ultimate goal of this book is to teach you the brainwave mastery that develops the Awakened Mind brainwave pattern found in people like swamis, yogis, healers, and spiritual teachers whose states of consciousness we emulate in many ways. Through guided exercises and specific meditations designed to elicit and develop this state, you can open the flow of information between the conscious, subconscious, and unconscious minds that characterizes an Awakened Mind and apply this to your own specific needs to achieve a high-performance mind.

When I sat down to write *The High-Performance Mind* ten years ago, I wanted to put all of the work that I had been developing down on paper so that I could offer people an effective training to produce the brainwaves of high performance. I had no idea at the time that I would actually be initiating a new field, and that what I was writing would eventually become, in the eyes of many, the primary foundational information on training higher states of consciousness.

This material is applicable to any and all situations where people want to train their minds to create higher states of consciousness, and readers that have used this information to develop a high-performance mind have come from backgrounds as diverse as a Chinese Kung Fu master looking to understand the physiological basis of his spiritual martial art to a stock market day trader wanting to sharpen her mental skills and intuition to increase her success in the stock market. After making this work public, I found that it was received and appreciated by a far broader spectrum of people than I had ever imagined, and, judging by the e-mails, it has touched people from all over the world; a reflexologist from Buenos Aires, a restaurant owner from London, a spiritual seeker from Poona, India, a management consultant from Sydney, a kindergarten teacher from Taipei, a classical pianist from Amsterdam, a psychiatrist from Kiel, Germany, a record producer from New York City, a CEO from Portland, a television executive from Washington, D.C., a healer from Mt. Shasta, California, and a Zen master from San Francisco are representative of the many people who have let me know that they have benefited from *The High-Performance Mind*. For this, I am humbled and grateful.

The book has been translated into Chinese, German, and Russian, and I have presented seminars in South America, Asia, Australia, Europe, and North America to large gatherings as diverse as The Broadcast Design Association's annual television design conference in Toronto, to the Sixth Professional Golfers Association of Europe

Teaching & Coaching Conference in Munich, where I spoke to one thousand golf pros about meditation and biofeedback and taught them how their game can be enhanced by playing with a high-performance mind. The variety of uses and applications of these concepts is limited only by the mind itself. Wherever you would like to improve your ability, increase your functioning, or gain greater mastery, developing a high-performance mind can help you.

I have presented this work at a variety of conferences and have been invited every year since this book was published to speak at the Futurehealth Winter Brain Meeting, a gathering of cutting-edge EEG scientists, practitioners, and researchers. Futurehealth's support of this work and invitation to teach on the "EEG Biofeedback/Neurofeedback Foundations Course" is testament to the recognition and acceptance this has gained in that community.

Out of all the different reasons people have been drawn to this work, the greatest pull has been spiritual. A craving for an understanding and experience of the ineffable is by far the major motivating factor in the majority of people seeking instruction and inspiration from this work. People from *any* spiritual tradition can and do relate to the material to help them move more closely toward the state that they desire for their heightened spiritual awareness.

Many people have found the "stand-alone" practice offered here helpful to access realms of universal wisdom and divine source. Whether these teachings are used as an adjunct to your spiritual practice or used as the sole modality for spiritual development, by practicing the exercises and meditations in the book, readers can begin to gain mastery of their own highest potential.

It is important when you follow the guided imagery exercises that you interpret the meaning of your meditation imagery yourself, rather than allow someone else to explain your experience to you. For example, one meditator's imagery was full of knives of every shape and size: paring knives, butcher knives, cleavers, serrated knives, pocketknives, and hunting knives. You can imagine the variety of interpretations one might give this image. When I asked him to tell me what he got out of the meditation, his answer was, "I found my *cutting edge!*" He was thrilled with himself about this self-discovery.

One crucial step in the brainwave training protocol is learning to master beta brainwaves, learning to turn off the "squirming of the worm in the brain," as Patanjali, the second-century B.C. author of the Yoga-Sutra, called our chattering mind. Today it might be called the list-maker, judge, critic, or constant commentator. I have had countless people thank me and tell me what a difference this simple but profound piece of information has made to them, not only in their meditations but at other times when they are trying to clear their minds of anxious, unwanted, or unnecessary thoughts.

One of my students told me the story of a friend of his who had done what he called "the great retreat"—three years, three months, three weeks, three days, and three hours of intensive meditation practice. About a week after he returned from his retreat, his friend gave him *The High-Performance Mind* to read. When he read the part on the importance of relaxing the tongue to turn off unwanted thoughts (page 86), he clapped his forehead in exasperation and said, "Oh no! If only someone had told me before I wasted three years!" You can cut off many years of "sitting in the cave, waiting for it to happen" by following some of these simple principles.

The House of Doors is a theta brainwave development exercise that takes us into the depths of the subconscious mind, for only in accessing and opening up these deep wells of inner resources can your mind begin to be awakened. In theta development work, many people find they get in touch with hidden and often previously inaccessible parts of themselves. As you will find, these meditations take you into your own subconscious and help you have a greater understanding of what might be hidden there, allowing you to get more deeply in touch with your inner wisdom, and, if appropriate, to begin to transform or heal any wounds or unresolved issues that you find.

In one self-healing meditation (page 131), a young woman discovered a dark unidentifiable shape. Following the instructions, she dialogued with it and found that it was an illness inside her that was speaking to her. In a recent gynecological checkup, she had been told she was fine, but in her theta state she got the message that she had cancer. Seeking a second opinion, she went through another battery of tests only to be told again that everything was negative.

Subsequent meditations revealed the same black mass in her subconscious, urging her again to seek medical assistance. The tests continued to "prove" that nothing was wrong. She eventually found a doctor who understood the power of meditation and was willing to do a simple exploratory surgical procedure to investigate her concern. The doctor did indeed find a mass of cancer, hidden in a location that the tests would not have found until it was "too late." She was very grateful when she told me that her theta meditation might have saved her life.

The contents of the subconscious are limitless and can be revealed to you in many ways. I asked one young woman to draw a picture of her meditation. She also had seen an image of a small dark round mass in her uterus. The picture she drew was of a seed planted in the earth. She had no idea what it might mean to her, and was actually a bit disappointed because she felt that she had nothing "exciting" in her subconscious—just a brown seed. When I ran into her again several months later, she was quite obviously pregnant and glowing. I asked her about the timing of her pregnancy in relationship to when she did the meditation. She blushingly told me that the baby had been conceived the night before the meditation—the seed in her image was her subconscious telling her she was pregnant!

The results of the many meditations in the book are as varied as the readers. In one meditation, a woman encountered the part of herself who had agoraphobia and had been afraid to ride in an elevator for the past twenty years. The day after she practiced the Self-Healing Meditation 2 (page 136), she told me that the phobia had completely disappeared. The CEO of a Fortune 100 company told me that after practicing Personal Creativity Meditation (page 172) for problem solving, he "figured out how not to have to fire fifty people," an action he was facing the following day.

A mother e-mailed, writing, "I initially bought the book for myself but found that you are also mentioning children in the meditation context. I tried the Purple Planet meditation (page 187) at bedtime with my seven-year-old daughter. She responded very well. Her description of what she saw on the purple planet was very vivid and detailed." She told of the impressive lesson her daughter had learned about "how to have real fun."

A suicide prevention hotline telephone worker uses Opening to Clarity (page 230) as a morning meditation to clear away his unresolved emotions. He told me of one specific instance where he was having extreme difficulty dealing with his bosses' anger, unpleasantness, and hostility. By practicing this meditation that focuses on these specific individuals, he was able to let go of the situation emotionally and come to a place of clarity. In addition, he "gained the insight that what those people needed was light and love." He told me, "My bitterness and animosity wasn't going to change them or improve me." So instead, he began "sending them light," and quickly found that "they started being less demanding and more generous in spirit."

So how do so many people have such a wide variety of successful experiences, often using the same meditation to very different ends? The answer is complex and will be best found by practicing the exercises in the book. Specifically designed content (words, colors, sensory imagery, and other specialized instructions) are combined with deep relaxation to help you develop and gain mastery of particular brainwave combinations. These brainwave states then allow you to access deep universal experiences of meditation and higher states of consciousness, individualize them to your own essential being and precise needs, and use your own particular deep inner resources, history, and emotions in a very personal and intimately meaningful way.

The call to this work started in me more than thirty years ago and continues to be strong and spiritually driven. My involvement in helping awaken people is a gift and a privilege for which I am very thankful. My greatest reward is mirrored in one of my readers, who, upon reading The High-Performance Mind, said, "After spending forty years seeking on a number of spiritual paths, this work has finally provided the fulfillment I was in search of."

YOU AND YOUR
BRAINWAVES

A high-performance mind is one that can enter at will the state of consciousness that is most beneficial and most desirable for any given circumstance. Once the aspiration of only a devoted few, mental, emotional, and spiritual development has taken the West by storm. In the final stages of the twentieth century, an abundance of developmental techniques drawn from many cultures, belief systems, and rationales has arisen. They address creativity, mental clarity, stress management, emotional and physical health, and personal spirituality. With the techniques described in this book, cultivating one's mind power has reached a new level that will carry us well into the twenty-first century.

Combining the science of brainwave measurement with meditative tools, my late mentor, Max Cade, and I developed a method for examining and developing optimum states of mind. This book is designed to help you achieve the specific brainwave patterns that are ideal for you.

While many people may desire the high-performance mind, they often believe that it is not possible to control their states of consciousness, except perhaps with years of training or meditation. They view this form of self-mastery as beyond the reach of normal individuals who do not want to spend hours every day in a contemplative practice. But the marriage of meditation and technology has created a new approach to mind mastery, one that involves learning the details of brainwave development for specific purposes.

The theories, exercises and techniques offered in this book will take the reader on a journey that ranges from taming out-of-control thoughts to experiencing pleasurable excursions into vivid sensory imagery to diving into the mind's deepest core. Once readers have learned the techniques for producing their optimum brainwave patterns, later chapters will show them how this mind power can be applied beneficially to their lives.

The high-performance mind possesses a potential for using optimum states of consciousness for greater creativity; self-healing; better general health, relaxation, and stress management; solving emotional problems; more productivity in the workplace; understanding and improving relationships; greater self-knowledge; spiritual development; and the nurturing of one's children's inner world. After reading the book, the reader should have a good understanding of his or her own states of consciousness and how to intentionally alter them for these specific purposes.

Practicing the exercises and techniques in this book over time can ultimately enable the reader to develop the brainwave pattern called the *awakened mind*. This state of mind is clearer, sharper, quicker, and more flexible than ordinary states. Thinking feels fluid rather than rigid. Emotions become more available and understandable, easier to work with and transform. Information flows more easily between the conscious, subconscious, and unconscious levels. Intuition, insight, and empathy increase and become more integrated into normal consciousness. With an awakened mind, it becomes easier to visualize and imagine, and to apply this increased imagination to one's creative processes in many areas. As a result of all this, one has a greater feeling of choice, freedom, and spiritual awareness.

THE LANGUAGE OF BRAINWAVES

To begin this journey, you must first understand the building blocks we will be working with. Your brain is producing electrical impulses all the time. These currents of electricity, or brainwaves, are measured in amplitude and frequency.

The *amplitude* is the *power* of the electrical impulse, measured in microvoltage.

The *frequency* is the *speed* of electrical undulations, measured in cycles per second (or hertz). The frequency determines the *category* of brainwave—*beta, alpha, theta,* and *delta.* The combination of these categories determines or underlies your state of consciousness at any given time.

Each state that you experience entails a symphony of brainwaves, with each frequency playing its own characteristic part. Out of these symphonies come the art of Picasso, the dance of Martha Graham, the architecture of Frank Lloyd Wright, and the theories of Einstein.

This finely woven, intricate interrelationship of brainwave frequencies delicately determines your state of consciousness. While you are rarely producing only one type of brainwave at a time, each category of brainwaves has its own qualities and characteristics. In mastering your mind, it helps to become familiar with the individual categories separately. This will assist you later in understanding their combinations. In the next four sections, I will describe a state of mind that will be familiar to you and connect it with the brainwave pattern it represents.

BETA BRAINWAVES
From Normal Thought to Panic

Thought after thought keep colliding in your brain. You can't stop them, and you can't seem to slow them down long enough to focus on just one of them. Your heart is racing, temples are pounding, breathing is rapid, and you can't think straight. Your mind seems out of control, and your *beta brainwaves* have run riot.

The brain usually produces beta brainwaves in its normal waking states, but excessive amounts of beta can lead to great psychological discomfort. When functioning at their best, beta waves are associated with logical thinking, concrete problem solving, and active external attention. Beta waves help us to consciously function in the world, but we also need to know how to master them so that we can use them effectively and not be at their mercy.

C. Maxwell Cade, grandfather of British EEG biofeedback, defines beta as "the normal waking rhythm of the brain associated with active thinking or active attention, focussing on the outside world or solving concrete problems. The strength of the signal is increased by anxiety and reduced by muscular activity."[1] Beta is associated with increased blood flow and increased metabolism, and is an active state with higher levels of cognitive processing, complex thinking, and decision making.

ALPHA BRAINWAVES
"I get lost in daydreams . . ."

Your eyes are probably closed. Uninvited flashes of imagery vividly dance across the visual screen of your mind, as real as if you were actually there. Your attention may jump from scenario to scenario, settling to follow one path in detail and then shifting, seemingly without warning or reason, to another. The array of images and other sensory input doesn't have to make sense. The outside world falls away and you are absorbed in reverie with your *alpha brainwaves*.

Alpha brainwaves are present during daydreaming, fantasizing, and visualization. They are also associated with a relaxed, detached awareness and with a receptive mind. Some people produce too much of these middle-range frequencies and might experience life in a haze of daydreaming and fantasy, perhaps even enjoying this escape from more ordinary reality.

The most common problem with alpha brainwaves is not having enough of them *in conjunction with other brainwaves*. Alpha provides the bridge between the conscious and subconscious mind. Without alpha, you will not remember your dreams when you awaken, even though you recall that they were strong, vivid, and meaningful. Without alpha, you will not recall your meditation even though you *know* you went to a great depth and had many insights. When alpha is missing, the link to the subconscious is broken.

Cade states that alpha is the brain rhythm with the "least apparent meaning unless the associated [brainwave] rhythms are also known."[2]

Alpha, however, was the first brainwave that people were able to learn to identify and control.

Joe Kamiya, the grandfather of EEG in the United States, discovered in his early groundbreaking research at the University of Chicago in the late 1950s that he could train individuals to discriminate the presence or absence of alpha. Continuing his research at the University of California, Kamiya trained people to speed or slow their alpha by giving them higher- or lower-pitched clicks as aural feedback, thus initiating EEG *biofeedback*. These simple but vital beginnings were the forerunners of the more complex investigations of Cade, who researched brainwave biofeedback for the development of higher levels of consciousness.

Kamiya's work was revolutionary in the sixties. The popular attitude that developed later unfortunately overemphasized the importance of alpha, hindering the progress of brainwave training and its potential for manipulating states of consciousness. Alpha eventually was portrayed as a kind of be-all and end-all of brainwaves. The general misconception was that if you could produce alpha, you were somehow truly succeeding to produce an "ultimate" state, whether that state be meditative, creative, or simply an altered one.

Many neurophysiologists, EEG technicians, and clinicians involved with the brain understood that some people could produce alpha easily, perhaps simply by closing their eyes or even while watching TV, while others could not. They concluded that the extraordinary magic and power attributed to alpha could not possibly be true. The unfortunate consequence of this situation was that EEG monitoring and feedback for consciousness was devalued and neglected. It took years for brainwaves to begin to regain their proper status as an important reflection of an individual's state of mind.

THETA BRAINWAVES
"It came to me out of the blue!"

You feel a kind of niggling in the back of your mind, a persistent but indefinable nagging that tells you there is something wrong but will go no further in defining what it is. You may experience this sensa-

BRAINWAVE PATTERNS

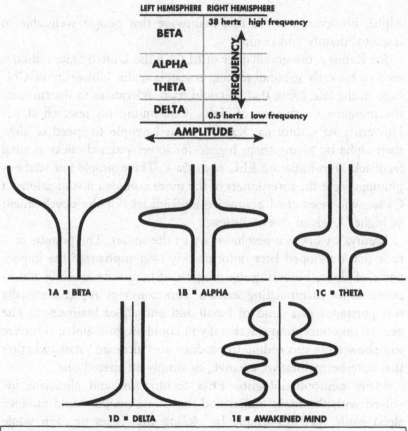

The drawings portray brainwaves measured in frequency and amplitude. The left hemisphere is shown on the left side of the drawing, the right hemisphere on the right side. Unlike traditional EEGs, the frequency is measured on the vertical axis with the highest frequencies at the top of the axis and the lowest at the bottom. The range of frequencies on the Mind Mirror EEG is from 38 cycles per second at the top down to 0.5 cycles per second at the bottom.

Amplitude is measured on the horizontal axis in microvolts. Amplitude measurements start at 0 in the center. The farther out to the left edge the line goes, the higher the amplitude in the corresponding frequency in the left hemisphere. The farther out to the right edge the line goes, the higher the amplitude in the corresponding frequencies in the right hemisphere.

From this you can see that Figure 1A portrays the strongest amplitude in the high frequency beta range of 38-14 Hz. (cycles per second).

Figure 1B illustrates the middle frequency range that is alpha (14-8 Hz.).

Figure 1C shows the lower frequency brainwaves of theta (8-4 Hz.).

Figure 1D illustrates the very lowest frequency brainwaves that make up delta (4-0.5 Hz.).

Figure 1E illustrates the combination of frequencies from beta, alpha, theta, and delta that unite to make up the awakened mind brainwave pattern.

The rest of the illustrations in the book are drawn using combinations of brainwaves from one or more of these four frequencies.

tion as something pushing at you from within, some kind of knowledge that wants to get out but is locked away deep within the recesses of your psyche. Sometimes you may even get close enough to it to feel an overwhelming sense of spiritual awakening, an inexplicable burst of creative insight, a cavernous sense of pain, or a dawning awareness of the possibilities of what might be hidden within your mind. When you experience these sensations, your *theta brainwaves* are trying to tell you something.

Theta brainwaves can be thought of as the subconscious, that part of our minds that forms a layer between the conscious and the unconscious. Theta holds a profusion of memories, sensations, and emotions. Though these experiences may be inaccessible to the conscious mind, they can nevertheless affect and even govern your attitudes, expectations, beliefs, and behaviors. The adult who cannot remember childhood abuse continues to seek out abusive people because the part of his or her mind that produces theta waves is driven to release that deeply held secret.

Theta brainwaves can also be the repository of suppressed creativity and inspiration. They are active during dreaming sleep and deep meditation, and they are particularly strong during so-called peak experiences, spiritual insight, and high-functioning brain states.

While theta is associated with creative activity, other brain rhythms are needed to bring this creativity into your conscious mind. A combination of additional frequencies is needed to access the profound potential of theta.

DELTA BRAINWAVES
"I feel as if I read people's minds.
I know what they are going to say
before they say it."

You know the telephone is going to ring before it rings. You know what your friend/lover/spouse/child is feeling because *you feel it too*. Sometimes you are even confused about whether the feeling you are experiencing belongs to you or to someone else. You desperately want to reach out to someone—to contact him or her—but you are afraid. Somehow you reach out with your mind and before you know it, the person contacts *you*. Your *delta brainwaves* have become your own personal radar, sending and receiving messages on an unconscious level.

Delta brainwaves make up the unconscious mind. Present during deep sleep, they are what is still turned on when the rest of your brainwaves turn off. Delta provides the restorative stages of sleep.

Delta may also be present in a waking state in combination with other waves. I like to think of delta as a kind of radar that seeks out and receives information on an instinctive level. People with high-amplitude delta waves are often intuitive and, through trial and error, have learned to trust their sixth sense, simply because it is so often correct. High-amplitude delta creates strong empathy. It is often present in large quantities in psychotherapists, healers, counselors, and others in the helping professions.

The ability to access delta may sound like a blessing, but for those who have a high degree of these brainwaves, especially people who don't understand what is happening to them, it may feel like a curse. I often see people who are overwhelmed by their receptivity to other people's feelings, needs, and thoughts. In some cases, they need to be taught to screen out the seemingly continual barrage of input they are picking up on an unconscious level. They need to learn to develop healthy boundaries and to determine the difference between which feelings are theirs and which rightfully belong to other people.

HOW BRAINWAVES COMBINE TO GIVE US OUR EXPERIENCE

We have now looked thoroughly at the four components of your brainwave pattern: beta, alpha, theta, and delta. Every state of consciousness that you are in at any time is a combination of these four categories. By examining how these categories interact and work together, we can learn much about what is happening in our minds and why certain experiences occur.

While the popular consensus of the seventies was that alpha held the answer for those seeking higher states of consciousness, in the nineties theta and delta have moved into the New Age scrutiny. Some current purveyors of meditation fads claim that theta and delta brainwaves are at least more important than beta or alpha, if not *the* answer to higher states. Others in the pop culture teach that there is a hierarchy of brainwaves, ranging, in order of importance, from beta to alpha to theta to delta. Others do not even address delta waves, considering them irrelevant or impossible to measure. The truth of the matter is that no one brainwave category is better than any other but that all work together in a variety of important ways to create a wide range of mental, emotional, and spiritual states.

Within the broad categories of beta, alpha, theta, and delta, there is actually a range of several frequencies. These vary from person to person, with each individual having his or her own *signature pattern,* which continues to resemble itself even though it is changing as our states of consciousness shift and flow. The different frequencies can be thought of as musical notes in a symphony. Sometimes the combination is melodious, smooth, and flowing. At other times, the discordance can become almost a cacophony. Our goal is to create a full, powerful, and harmonious symphony of brainwaves. As with music, mastery of the notes is our key, as well as learning how to build them into meaningful combinations that accomplish what we want.

LEFT HEMISPHERE VERSUS RIGHT HEMISPHERE

Popular conception often has it that the left hemisphere of the brain is the logical or thinking side, the right hemisphere of the brain is the creative or artistic side, and we are usually using either one or the other. In reality, the brain is an integrated whole with both sides in constant use. The left hemisphere function is more linear and detail oriented and the right is more spatially oriented and concerned with the big picture, but both functions are needed for you to perform fully in the world. Optimally, both hemispheres operate symmetrically, but an asymmetrical pattern frequently occurs. Meditation and brainwave training will improve the balance between hemispheres.

You think in both left and right hemispheres and you create in both left and right hemispheres. The shift that occurs to bring people to a fuller, more creative state is not a movement from left to right. Rather it could be viewed as an expansion from up to down—beginning with the thinking waves of beta and adding the creative aspects of alpha, theta, and delta in both hemispheres simultaneously.

THE AWAKENED MIND

"Ah-ha! I got it. Everything makes sense now. For a few moments I completely and totally understood. Why can't I be like this all of the time?"

Exhilarated, you know that you know! You have a sense of understanding on all levels. Perhaps this sudden awareness is leading you to thoughts and acts of extraordinary creativity. Your intuitive insight into old problems makes them seem simple and easy to handle, possibly even insignificant. "Why didn't I think of that before?" or "It's so obvious!" may be your response to what at one time seemed like an unanswerable question. Or perhaps your experience takes the form of increased spiritual awareness. You have a feeling of greater knowledge of the universe—maybe even a sensation of being surrounded in light. You are experiencing an awakened mind.

The *awakened mind brainwave pattern* is a combination of all four of the above brainwaves—*beta, alpha, theta,* and *delta* simultaneously. When you produce all of these frequencies together in the right proportion and relationship with one another, you experience the intuitive, empathetic radar of the delta waves; the creative inspiration, personal insight, and spiritual awareness of the theta waves; the bridging capacity and relaxed, detached, awareness of the alpha waves; and the ability of the beta waves to consciously process thought, *all at the same time!*

DISCOVERY OF THE
AWAKENED MIND

The late C. Maxwell Cade was a distinguished British psychobiologist and biophysicist and one of the few nonmedical members of the prestigious Royal Society of Medicine. Cade grew up with a tradition of meditation and Zen training. As his scientific abilities developed, he became interested in finding out whether the brainwaves of people who were experiencing higher states of consciousness could be measured. If so, would he be able to better understand what was going on in the minds of these select people? Would this help him teach others to more easily enter these states? He was already teaching a variety of forms of meditation techniques with the aid of electrical skin-resistance biofeedback meters. Finding a way to measure people's brainwaves was the next logical step.

To this end, Cade, along with electronics expert Geoffrey Blundell, developed an electroencephalograph (EEG) designed specifically to accomplish this goal. They named their invention the Mind Mirror. It differed from traditional EEGs in that it used spectral analysis to simultaneously measure eleven different frequencies in each hemisphere of the brain.

Max took the Mind Mirror to a variety of swamis, yogis, healers, and advanced meditators whose states of consciousness people emulated. When he measured them, he found the common thread for which he had been searching. He observed a recurring pattern of brainwaves in both hemispheres that he called the awakened mind. Cade continued to find confirmation of this lucid state in the highly evolved minds that he studied, and learned how to help his students develop it.

When I returned to the United States in 1981, after eight years of studying and working with Max Cade in London, I began conducting my own investigations. I found that the awakened mind pattern was produced at the moment of creative inspiration, regardless of a person's spiritual dogma, belief, or tradition. The musician composing, the choreographer creating a dance, and the artist painting all produced this combination at times of peak creativity. The mathematician solving a difficult equation and the scientist in the midst of an experiment also display this brainwave pattern. The CEO making decisions in the boardroom and the homemaker keeping an exquisitely appointed house both may produce this pattern.

For most people, the awakened mind state comes only sporadically, in flares of inspiration, perhaps only for a second or so. You have probably experienced at least one, if not many, of those moments of clarity and lucidity in your own life. Perhaps there is a certain activity that you enjoy that helps you create this mental state easily. For others, the awakened mind provides a rare and treasured dip into a desired but rarely attainable high. Whatever your current state of consciousness and your level of control over it, you can begin now to learn to master your own brainwaves and awaken your mind.

In this book we will ultimately dissect the awakened mind brainwave pattern, learn what makes it work, and look at building blocks for developing it. But first, let us look at a variety of mental starting places and the brainwaves that accompany them.

COMMON STATES OF CONSCIOUSNESS

Figures 2A, B, and C illustrate the brainwave patterns for states of consciousness that you might experience during your waking hours. Because you may have them frequently, these patterns are likely to contain the building blocks with which you start, that will help you to learn to master other, more advanced states. As you learn mastery techniques, these basic states provide a language of consciousness through which you can understand your experiences.

In a waking state, most of us produce what I call splayed beta (Figure 2A). If you refer to the picture of the awakened mind pattern (Figure 2E), you can see that the beta displayed there is at a much lower frequency. Splayed beta can be experienced as an overabundance of mental activity—the parts of our mind that are busy constantly commenting, planning, judging, or criticizing. Some call it

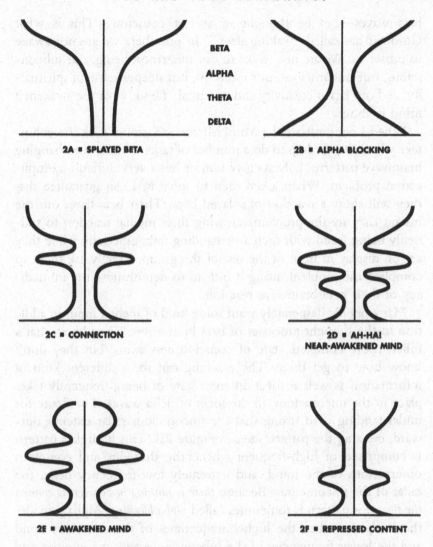

2A ▪ SPLAYED BETA

2B ▪ ALPHA BLOCKING

2C ▪ CONNECTION

2D ▪ AH-HA!
NEAR-AWAKENED MIND

2E ▪ AWAKENED MIND

2F ▪ REPRESSED CONTENT

the committee—that part of us that continually talks to ourselves—or the list maker. Splayed beta can also be experienced as panic—worry, anxiety, or just a vague sense of uneasiness and tension.

Many people can successfully operate out in the world producing only, or at least primarily, this brainwave pattern. Apart from being a somewhat uncomfortable mental state, splayed beta is not a fully conscious state, either. How interesting that the conscious mind—

beta waves—can be thought of as not conscious. This is what Gurdjieff has called "waking sleep." In pure beta we are not awake to ourselves. We are not awake to our innermost being, our subconscious, our unconscious, our intuition, our deeper sense of spirituality, and our latent creativity and potential. This is what the awakened mind is about.

When I demonstrate the Mind Mirror in a group, I ask the volunteer who is hooked up to do a number of tasks to show the changing brainwave patterns. I always give him or her a very difficult multiplication problem. When asked then to solve it, I can guarantee that they will show a reaction of splayed beta. Their beta flares out the instant they *see* the problem, showing their mental reaction to suddenly being faced with such a demanding task, especially while they are on display in front of the rest of the group. I rarely ask them to complete the problem, using it instead to demonstrate the immediacy of their beta brainwave reaction.

Many people desperately want some kind of mental input in addition to the thought processes of beta brainwaves. They know that a fuller, more complete state of consciousness exists; but they don't know how to get there. The reaching out for a different kind of information as well as for a different state of being frequently takes place in the unconscious, in the form of delta waves. The desire for understanding is so strong that the unconscious radar extends outward, creating the pattern seen in Figure 2B. This hourglass pattern is comprised of high-frequency beta, the thoughts and conscious observations of the mind, and extremely low-frequency delta, the radar of the unconscious. Because *there is nothing in between to connect the two*, this pattern is sometimes called *alpha blocking*. Alpha provides the bridge between the higher frequencies of the conscious mind and the lower frequencies of the subconscious and unconscious and it is significantly missing.

The person exhibiting this alpha blocking pattern really wants to know what is going on. Both his conscious and his unconscious are seeking input and information, but neither is aware of the activity of the other.

I frequently see this pattern in clients when they are first hooked up to the Mind Mirror. Next to splayed beta, this is the second most common pattern that I see in people in a normal waking state—especially when they are in the unfamiliar position of having their brainwaves monitored.

When an individual is wired up to the Mind Mirror for the first time, he or she is usually thinking fairly strongly, so I expect to see strong beta. In fact, I would be concerned if I didn't. The appearance of delta, however, indicates a deep level of interest in what we are doing.

When I get to know my client's brainwaves, I can often tell if she is losing interest in what I am saying when I see her normally strong delta begin to withdraw. At that time I can shift the energy of what I am saying and bring her delta level back up again. In other words, I can keep her unconscious interest level high—a kind of active non-verbal engagement in what is taking place. During intense therapy sessions, the delta is often very strong as she searches for insight with her unconscious mind.

I usually keep my clients hooked up to the Mind Mirror during every session for this reason. I get valuable information about what is taking place inside their minds by watching their shifting brainwave patterns, often seeing changes in consciousness of which they themselves are unaware.

If you are creating this beta and delta pattern, you may experience it as a normal thinking state, but with an added sense of underlying searching—a desire to reach out to or into what you are involved in, to truly know and fathom the meaning of the experience you are having. Remember that delta represents the *unconscious,* so that your searching is being done on a level that is not self-aware. Alpha blocking may also serve to protect people from thoughts they just can't consciously handle right now.

Alpha blocking is not always stable but may be accompanied by continuous or short bursts of alpha waves (Figure 2C). When these occur, you are beginning to have a connection between your conscious and your unconscious minds. The more frequent and longer the flares of alpha, the stronger that connection. When this state occurs, an individual might feel more aware of his intuition and empathy and be more in touch with his deeper feelings and needs. Although he is probably experiencing more vivid imagery and a better use of imagination than he had previously, he may sense that something is still missing.

Figure 2D shows what is commonly called the "ah-ha" experience. For a brief period or even longer, the person experiencing this state will have found that missing piece. You may not know exactly *what* you understand, but you will still have that almost physical thrill running through you that tells you that you *do* understand. And often just as quickly, it is gone—and you wonder what it was that you just experienced.

These brief near-awakened mind patterns can happen frequently and are often characterized as very brief (split-second) flares or flashes of insight. If they are strong enough and long enough, the mind has enough time for the information to flow freely between the conscious, subconscious, and unconscious minds so that the content of the insight penetrates all levels and is able to be retained in conscious memory after the brief brainwave flare has ended. Just as frequently, the flare of beta, alpha, theta, and delta is just too brief for this to happen. You are left with the sense that you had it but just lost it. Often if you stop, close your eyes, and turn inward for a minute soon after the passing of this flare, you can recapture the content of it.

When describing Figure 2D, I called it a near-awakened mind. Why "near" and not the real thing? To see the difference, look at Figure 2E. Notice that the beta in the awakened mind curves inward at the top. If you are experiencing an insight while in the state of consciousness represented in Figure 2D, the splayed beta will also give you a sense of distraction, or excess thoughts. You could alternately experience these accompanying symptoms as a lack of focus, overexcitement, anxiety, attempts at analysis, or some other form of internal preoccupation. I have seen the pattern in Figure 2D many times in very successful executives, who have the insight but just can't let go of the anxiety.

When the awakened mind of Figure 2E is experienced even briefly, the sensation is one of focus, clarity, and unity, with no extraneous or unnecessary thoughts or distractions.

Figure 2F shows the brainwave pattern of beta, theta, and delta, but it is significant that the alpha is missing. An individual in this state might experience some discomfort, especially if the theta is strong. The theta content of the subconscious mind may be negative or positive. No matter what that content is, you may experience it as pushing at you from underneath. With no alpha to bridge the gap to your beta waves, there is no way for you to consciously know or understand what it is that is pushing at you. Because we are sometimes strongly influenced by unknown contents of the subconscious, our behavior can be affected without us knowing why. When we are able to add alpha to our brainwave patterns, we often find the explanation for certain actions and attitudes we have exhibited. This gives us a greater level of choice over them.

The state represented by Figure 2F may be experienced by someone who has repressed psychological material stuck in his or her subconscious. It may be manifested as a driving inner force. This pattern can be accompanied by high levels of frustration—a creative urge or impulse trying to be expressed but not making it to conscious manifestation. This individual needs alpha development badly.

One client of mine who exhibited this pattern appeared, on the outside, to be the healthiest and happiest of men. He was successful in his work, content in his relationship, and emotionally comfortable and clear with his past. When I saw this pattern, however, I began asking questions and trying to ascertain what might be trapped in his subconscious. He quickly revealed that he had not been happy in his relationship with God for some time. The need for some kind of spiritual quest was welling up from within, filling him with a pressure and dissatisfaction that he didn't understand and didn't know what to do about. When I used his brainwave pattern to explain what was happening, he was greatly relieved and, with my help, began learning techniques to make his subconscious needs more conscious.

Another client with a similar pattern was a victim of childhood abuse. Her memories were deeply repressed, and her state of mind was very painful. With many sessions of work, she could retain enough alpha to begin to confront and heal what had happened to her.

HOW DO I CHANGE?

So far, we have talked about states of consciousness and the brainwaves behind them in a conscious waking state. By now you may have a better idea of the brainwave states you would like to be able to produce. You may also have some idea of where you are now —what brainwaves you are producing much of the time. So how do you develop the skills to shift your state of consciousness to be where you want to be whenever you want to?

DO YOUR WORK INSIDE FIRST

Brainwaves are much easier to learn to control and develop when we are in a state of deep relaxation with our eyes closed. The mid-frequency waves, alpha and theta, are immediately more accessible as soon as the individual turns inside. We ultimately want to be able to produce these brainwaves when we are relating to the outside world as well.

But first, let us look at how we already may experience some of these inner states. Those of you who practice some form of meditation may be very accustomed to your inner world. But whether we meditate or not, we all have inner states with which we are familiar.

Sometimes when we close our eyes and go inside, nothing changes. We are so preoccupied with our thoughts that our normal, open-eyed, splayed-beta state remains the same. You may be familiar with this experience at times when you would *like* to relax and perhaps get something off your mind but are so absorbed with it that the thoughts keep right on coming. Nothing you do seems to be able to change your mental state. Refer back to figures 2A or 2B to see what your brainwaves look like. If your goal is mental stillness, clarity, and relaxation, you may find this state uncomfortable. Absorbed in the contents of your beta waves, you may experience too much static—too much intensity and distraction. If you are preoccupied with thoughts that are negative or worrisome, you may feel anxious or tense. Even if your mind is filled with positive, exciting, or creatively challenging thoughts, you may nonetheless feel agitated and uncomfortable because you can't get all these thoughts out of your mind.

The ultimate goal is to learn to master one's mental states, regardless of their content. It is just as important to learn to control the positive as the negative.

"But I don't have a long time to meditate. How can I do the work inside effectively in a short period of time? I want to get rid of these restless thoughts."

What happens when we have all the good intention in the world to practice meditation, but somehow something always gets in the way —work, the kids, our relationships, our hobbies and sports, the shopping, the cooking, the cleaning . . . ? Do we have to give up on the idea of self-mastery of internal states, of developing the high-performance mind? The answer is a resounding no.

Even a few minutes of meditation, or going inside, as infrequently as once a week can have an impact on your brainwaves and therefore your experience of life. Many people feel they have to dedicate as much as an hour a day to receive any benefit from meditation or relaxation. They may try it for a few days or even a few weeks. But their determination soon flags because they simply do not have that much consistent free time in their busy schedules—or they have not yet decided inside themselves that this is an important enough priority to rearrange their busy schedules.

At first, perhaps, guilt sets in. They make a few more efforts. Then the guilt becomes uncomfortable, so they decide that meditation is not for them. They put the practice of learning to master their states of consciousness into their past instead of their present.

What these people don't understand is that *any* time spent in inner reflection, contemplation, or relaxation is beneficial. If they give even a little time to this experience, they will feel the rewards, which will then encourage them to repeat it.

THE BRAINWAVES OF EMOTIONS

One of the most frequent questions I am asked is, "Are there specific brainwaves for love, anger, hate, joy, fear, and so forth?" The answer is not easy, because each of us has his or her own unique brainwave pattern for each emotion.

For example, one person's anger might involve an increase in amplitude of all frequencies—a burning rage causing large amounts of electrical activity to be generated in the brain. Another person might experience a dramatic reduction in the amplitude of all frequencies as his anger causes him to withdraw. The anger of a third person might generate one specific frequency of brainwave, such as a burst of beta waves, as out-of-control thoughts kick in, or of theta waves, as his rage stimulates a memory buried in his subconscious.

Your brainwaves might react differently to different kinds of anger, or they may respond to all types of anger in the same way. If I had the opportunity to watch your brainwave patterns on the Mind Mirror enough times, I would eventually be able to recognize when you were angry by watching your brainwaves alone—without even looking at the expression on your face—because I would come to know your particular patterns. But again, there is no generic pattern that is the same for everyone.

I am not saying that you can always control your anger by controlling your brainwaves. If there is something that makes you angry that you should have dealt with ten years ago, you may still need to deal with it. The benefit of changing your brainwaves is that this can enable you to alter your *experience* of the anger, how you manifest it, and how you handle it.

How, then, can you use brainwave mastery to temper your anger? The next time you get angry, really *notice* what you are experiencing mentally. Use the descriptions of the different categories of brainwaves and the different combinations of those categories found throughout this book to help you begin to understand what brainwaves you are producing and what your personal patterns are. Then use the various forms of brainwave mastery and self-control of internal states that you are learning in order to change those brainwaves into ones more in keeping with what you would like to have at the time.

You will soon come to recognize when you are not where you want to be and when you are where you want to be. Gaining the ability to intentionally change an undesired state to a desired one takes a little more time and effort. This process does not happen all at once but is a gradual and evolving one that unfolds as you practice. And the outcome is well worth the effort.

Before we close this chapter, I would like you to try a short meditation. After you have read through these instructions, sit comfortably where you can be relaxed and undisturbed for about ten minutes.

With your eyes closed, allow your mind to clear of all thoughts. Spend a few minutes just focusing on your breathing and relaxing.

Mentally travel through your whole body, relaxing each part in turn. Starting with the muscles of your face, neck, and shoulders, relax all the way down to your feet and toes.

Allow yourself to experience your internal stillness, silence, and serenity.

From a place of deep peace and relaxation, allow yourself to find an image or a symbol which represents where you are in your life right now.

You may see a color or a visual image, feel a physical sensation, hear a sound or even a voice in your head, or just have a sense of knowing where you are in your personal journey.

Bring that experience back with you when you reawaken from your meditation.

Make a note of what your symbol was, so that you can remember it as you move forward through the book. You may wish to check back, over time, to see how you relate to this symbol—whether your path has remained the same or has changed, and how you are developing on it.

Cultivating a high-performance mind involves many facets. The rewards can be great, even from the beginning. You now have the basic language with which to continue to learn mastery of your beta, alpha, theta, and delta brainwaves, ultimately combining them to give you access to your own optimum states of consciousness.

1. C. Maxwell Cade and Nona Coxhead, *The Awakened Mind* (New York: Delacorte Press/Eleanor Friede, 1979), 24.
2. Ibid.

THE BODY
CONNECTION

I t is essential at this point that we look at the role your body plays
in altering states of consciousness for the high-performance mind.

Biofeedback is often used to help an individual learn to control
body states. We will now see how biofeedback can be used to help
you learn to control mental states as well. Biofeedback is simply the
feeding back of one's biology. You cannot consciously change something of which you are unaware. If you can be made to be aware of a
certain body function, then you can learn to alter it.

Biofeedback instruments measure some function in the body, giving you information about it. Armed with that information, you
have the power to then change it in some way. The most common
example of a household biofeedback instrument is a mirror. You
look at the mirror which feeds back to you your appearance. You
can then comb your hair or make some other change that will be
noted by the mirror, so you know that you have improved and look
more the way you would like to look.

Another familiar biofeedback machine is the bathroom scale. Although scales are often not used in the most effective way, it is
possible to lose weight through biofeedback by simply weighing
yourself each morning and having the intention to move the needle
on the scale down.

If you have a watch with a second hand on it, you can learn to
raise or lower your heart rate taking your pulse and counting the
number of beats for a few minute. Using fifteen- to thirty-second

increments, count your heart beats. You can intentionally slow or speed up your heart while you continue to look at the watch and count. Biofeedback is *not* a thinking process. Just allow the change to take place within you. Once you have made the decision with your beta waves, let your lower frequency alpha and theta waves take over to effect the change.

There is a biofeedback instrument for almost every body function that can be measured. Electrical skin resistance meters (ESRs) or galvanic skin response meters (GSRs) feed back the amount of arousal or relaxation of the nervous system and are used for teaching relaxation and stress management. Electromyographs (EMGs) feed back muscle tension and can be used for both teaching relaxation and retraining damaged muscles. EEGs, of course, feed back brainwaves, and the right kind of EEG can be used for training brainwaves. There is even a biofeedback instrument that you swallow that feeds back the amount of stomach acidity!

You will probably not be using a biofeedback instrument while practicing the exercises in this book. However, this chapter will show you that there are ways to receive biofeedback directly from your body. Every time you become consciously aware of a change that takes place inside you while you are practicing the exercises, note the sensation. This feeling or experience will become your biofeedback and tell you where you are. When you learn to recognize the internal language of your body's biofeedback, you can use each experience as a landmark to feed your brainwave state back to you and help you deepen it. Biofeedback is one of the tools that help you reenter desired states at will and remain in them longer.

Learning to manipulate brainwaves to develop deeper states of consciousness and the high-performance mind is much easier when you are relaxed than when you are tense or aroused. Try gritting your teeth, clenching your jaw, making a fist, and ordering yourself to *"visualize!"* and you will see what I mean. Meditation means mastering the ability to relax the body as well as the mind.

AROUSAL IS THE OPPOSITE
OF RELAXATION

The *body* does not necessarily differentiate between different causes of arousal, such as worry, excitement, fear, anger, or exhilaration. These all might result in similar body sensations. You can usually note, however, a marked difference between the body sensations of arousal, such as increased heart rate, faster breathing, anxiety, nervousness, or panic, and the body sensations of relaxation, such as tranquility, calmness, lightness, or serenity.

In the first situation the body's sympathetic nervous system is turned on, and regardless of the reason, the result is arousal. In the second, the body's parasympathetic nervous system is activated, and the result is relaxation.

To see what arousal feels like, try these exercises:

1. Hyperventilate by breathing heavily for a few seconds. (Please be careful not to overdo it. If you start to feel faint, stop immediately.)
   Stop. Close your eyes. Notice what your body feels like. Make a mental note of all of the sensations.

2. Run in place for a minute or two.

3. Think about something *very* upsetting.

4. Think about something *very* exciting.

After each one of these, stop and notice what is happening inside your body.

For some people, it is more difficult to arouse the body quickly while doing the last two exercises, because they involve *thinking* about emotional states instead of doing something physical. Heavy physical activity often causes a more immediate reaction of arousal. It is often easier, moreover, to return to a state of relaxation when the arousal was caused by physical activity than when the arousal was caused by mental activity. Arousing the body by imaging certain emotional states may take a little longer, and once you have succeeded, this type of arousal is often harder to get rid of because of the many accompanying physical changes.

When you are aroused in this way, you are activating what is called the *fight-or-flight response*. Certain physiological changes are likely, including increased oxygen consumption, heart rate, blood pressure, and muscular tension. You will also experience a higher level of adrenaline and cortisone in your bloodstream.

When you are aroused, your blood supply tends to move away from the periphery of your body and into the major muscle areas. This is why you might get cold feet when faced with a frightening or exciting event. When your body undergoes these changes, *it is prepared for action!* That action may be running, fighting, or using extraordinary physical abilities in some other way. Under extreme pressure or fear for a loved one's life, people have been known to exhibit amazing feats of strength, such as lifting a car or moving other large objects.

The act of running or fighting eventually helps the body return to its normal state of balance. But when such action doesn't take place, many people are left with an increased heart rate, high blood pressure, adrenaline, and tension, which put their bodies at greatly increased risk for stress-related diseases. The fight-or-flight response first becomes habitual, then dangerous.

Taking the state of arousal to an extreme, some people become addicted to stress, using the outpouring of adrenaline and other physical changes to help them meet their deadlines. They may find themselves lacking the energy to create without the additional push they get from their stress responses. When someone needs stress to feel excited and alive, it can become like a drug. Those in high-pressure jobs—the stockbroker, the writer on a deadline, the emergency-room worker, the high-powered executive—may routinely have these experiences. Even without a high-stress situation, we can create this type of pressure for ourselves simply by leaving tasks until the last minute, refusing to delegate responsibility properly, over-scheduling our time, and expecting too much of ourselves. Many people who continually rush, push, and live life in a state of tension have no idea that that is what they are doing. You may have actually met the person who grits his or her teeth when you say, "Relax," and snaps back, "I *am* relaxed!"

One way to handle the buildup of stress in the body is to consciously create the action needed to reverse it. This isn't always possible, however. In the middle of a disagreement with your boss, you cannot just excuse yourself to go run around the block. The alternative of fighting with him is even less acceptable. Even throwing yourself into activity to work off the stress is not always an available or appropriate answer to a charged nervous system.

The answer is to be able to *consciously* deactivate the fight-or-flight response and activate the *relaxation response.*

It is hard for most people to learn to alter their brainwave states while experiencing this fight-or-flight condition of physical arousal. The feelings and physical sensations you experienced during and after the arousal exercises represent your biofeedback. If you can remember how this felt and identify the next time you feel these sensations in your day-to-day life, you will know when you are in a state of physiological arousal. The relaxation training that follows will teach you how to begin to master these states of arousal and let go of them.

To see what relaxation feels like, try this (read the exercise all the way through first, as usual):

Close your eyes and exhale deeply. Let your shoulders drop. Rotate your head gently and loosely until you find a comfortable balanced position for your head, neck, and shoulders. Let your jaw relax and hang loose. Relax your lips, tongue, and throat. Exhale deeply again and let go. Continue to breathe easily, slowly, evenly, and deeply for one or two minutes . . .

Stop. Notice what your body feels like. Make a mental note of all of the sensations. Compare these to the sensations you noted when you did the arousal exercises.

This is your *biofeedback.* Whenever you feel these same sensations of release and restfulness, you will know that you are beginning to activate *relaxation.*

Unlike the physiological changes you experienced while in a state of arousal, now your oxygen consumption, heart rate, blood pressure, and muscular tension are decreasing. Less adrenaline and cortisone are being pumped into your bloodstream, and you will eventually feel an increase in the temperature of your hands and feet.

Try these simple exercises a number of times until you can instantly and easily notice the difference between the physiological sensations associated with arousal and relaxation. You can use these feelings and perceptions as your biofeedback mechanism to give you immediate information on which state your body is experiencing and how intensely.

''WHICH STATE IS BETTER?''

Arousal is not inherently better than relaxation, or vice versa. Both states are important *at certain times.* What is optimum is *to be able to choose the level of relaxation or arousal that you want and to be able to produce that at will.* Once again we are back to self-mastery and choice.

Don't get the mistaken idea that arousal, tension, or quick response to stimuli from the outside world is bad. You can be too laid back just as you can be too hyped up. Think about the individual (we have all met him) who is lazily crossing the busy road. When he sees the car bearing down on him, the correct response is *not* to continue to amble along.

Perhaps you have had an experience with the person who is late for something when being on time is vital—an important meeting or a plane flight, for example. The individual, even when he knows he is late, just pokes along, unable to get himself moving faster, until he eventually arrives after the meeting has started or the flight has left. He doesn't even realize that he is not taking charge of his activation response and that he could live his life at a faster pace, yet still stay healthy and stress free.

We need to be able to activate our nervous systems quickly and vigorously when appropriate and *when we choose to.* We need also to be able to deactivate our nervous systems and activate the relaxation response just as quickly and intentionally. Practicing the conscious movement between the two extremes is an excellent way to learn how to have that choice in times of necessity.

The following is a deep relaxation exercise. There are many techniques for relaxing the body. The method I have put together here has been very effective when practiced in a seminar or individual brainwave training or biofeedback session. You have a number of options regarding how you might wish to use this and all of the meditations and exercises in this book.

OPTIONS FOR PRACTICING THE EXERCISES IN THIS BOOK

ONE

Read through the exercise or meditation from beginning to end. Make a mental note of the general idea of what you hope to accomplish during the meditation. Don't try to remember this goal word for word, but rather get a *feel* for it. If you understand and deeply internalize the instructions, you won't have to activate your beta waves to tell yourself what to do. If it is appropriate, you may want to pick out key images, or stepping-stones, to help you remember the order of the instructions. Then close the book, close your eyes, and take the appropriate position—either sitting or lying. Allow yourself to re-create inside what you have just read. Try not to talk to yourself or use words to instruct yourself—just allow the meditation to happen.

TWO

You may wish to read the meditation verbatim into a cassette recorder and then play it back to yourself. Having the voice gently droning in the background gives you something to lean against that often lets you go deeper inside yourself. Be sure to speak slowly and clearly, with a relaxed tone of voice. Also, allow yourself many long pauses with no instructions, so that you can have plenty of time to feel the experience inside. Initially, some people are disturbed by the sound of their own voice. This should pass after you have listened to the recording a few times.

THREE

Have a friend read the meditation to you. Be sure he or she follows the same instructions to speak with a relaxed voice and allow you ample quiet time. You may even wish to start a small group meeting, taking turns while one person reads and the rest are meditating.

FOUR

I have made professional cassette tapes of many of these meditations. To purchase these, see information in the back of the book.

WHAT POSITION SHOULD I BE IN?

For most of the exercises in the book, you will want to be in a position that will allow you to achieve deep physical relaxation. This may be sitting on a cushion cross-legged or with legs outstretched, or sitting in a chair, or lying on a flat surface. (Be sure you are lying symmetrically on your back, with your spine in alignment.) Reclining chairs are also good to meditate in. If lying down puts you to sleep, sit up instead. If sitting, have something comfortable to lean against, but keep your spine as straight as possible. Often people like to sit on the floor leaning back gently against a wall. If this puts you to sleep, sit up straight without leaning against anything.

Deep relaxation will help you enter the mental state in which activating the lower-frequency brainwaves of alpha and theta is a much easier and more natural phenomenon. Practice the following relaxation lying down, if possible. It is better to avoid lying on a bed, simply because the body has a deeply ingrained memory of using the bed for sleep and will tend to send you in that direction as soon as you begin to go deep. Lying on the floor is just unfamiliar enough and just uncomfortable enough to keep you awake. To prevent back strain, you may wish to place a pillow or rolled-up towel under your knees to raise them slightly. A pillow under your head or neck might also be helpful.

Make sure the room is the right temperature. If necessary, fault on the side of warmth. Lying in a room that is too cold is not conducive to relaxation and may even increase your tension as your body fights the cold. Don't just hike up the heat, since too much warmth might also give you a tendency to fall asleep. A light blanket will keep your body warm during relaxation.

Timing is an important factor in relaxation. As long as you can stay present and maintain conscious awareness, the slower you go, the better. When reading these meditations into a tape recorder or giving them to another person, you may find the need to speak more slowly than you would anticipate. When I lead meditations and relaxations, my voice naturally pauses at specific places—not just between sentences. I have marked those pauses with a code:

(*)   = 2 to 5 seconds
(**)   = 5 to 10 seconds
(***) = 10 to 30 seconds

With practice, you will find your own sense of timing and specific needs.

DEEP RELAXATION

The next thirty minutes is a time for yourself,
a journey inside to calm, soothe, heal, and relax
your body, mind, and spirit. (**)

Begin by closing your eyes.
Allow your mind to clear of all thoughts, (*)
and focus on your breathing. (**)

Breathing easily and deeply . . . . (*)
easily and deeply . . . (*)
breathing relaxation into your body, (*)

and breathing away any tension, (**)
breathing relaxation into your mind (*)
and breathing away any thoughts. (**)

Very gently begin to withdraw yourself from the outside environment
    . . . (**)
Withdraw yourself from your surroundings . . . (**)
Withdraw yourself from any remaining thoughts . . . (**)
Withdraw yourself into yourself . . . (**)
Into your own silence . . . (**)
Into your own serenity . . . (**)
Into your own peace. (**)
And relax.

(***)

Allow the muscles of your face to relax . . . (**)
your forehead, (*) the muscles around your eyes, (*) the muscles behind
    your eyes . . . (*)
your lips, (*) tongue, (*) throat, (*) and jaws . . . (*)
. . . all deeply relaxed. (**)

Allow the relaxation to flow down through your neck . . . (*) into your
    shoulders. (**)
Allow it to flow down both arms . . . (*) all the way to your fingertips.
    (**)
Allow the relaxation to flow into your chest . . . (*)
and down your back and spine . . . (*)
. . . allowing the muscles of your back to just let go. (**)
You let go . . . (**) . . . you let go . . . (**) . . . you just (*) . . .
    let (*) . . . go . . .

(***)

Take the relaxation down through your torso. (*)
And allow it to go deep into your stomach . . . (*)
right into the very center of your body . . . (*)
right into the very center of your being . . . (*)

Allow the relaxation to flow down through your hips and pelvis (*)
and down through both legs . . . (*)
all the way to your feet and toes . . . (**)

. . . so that deep within yourself, you can visualize and experience your
whole body as completely relaxed. (**)
Deep within yourself, you can visualize and experience your mind as
quiet and still . . . (*)
. . . very still. (**)
Deep within yourself, you can visualize and experience your emotions as
calm and clear . . . (**)
. . . and your spirit as peaceful. (**)
Deep within yourself, you can visualize and experience your body, (*)
mind, (*) emotions, (*) and spirit (*) in harmony. (***)
Experience the relaxation . . .

(pause may be several minutes)

Very gently allow yourself to begin to find a closure for your meditation
. . . (**)
taking all of the time that you need to come to completion inside. (***)
Find an image, (*) symbol, (*) word, (*) or phrase (*) that describes how
you are feeling right now. (***)

Begin to allow yourself to return . . . (**)
back to the outside space. (**)
Allow yourself to reawaken and return . . . (*)
. . . feeling alert and refreshed. (**)

Take several deep breaths (*)
and allow yourself to stretch—beginning with your fingers and toes. (*)
Take a full-body stretch, (***)
and allow yourself to return to a sitting position. (***)

Before you resume your normal activity, take a few moments to reflect on
your experience. Remember your image, symbol, or words at the end.
Remember what it felt like to be so deeply relaxed, and once again
relate that feeling to the representative symbol or words you chose.

■ 33 ■

The next time you want to relax, recall this cue. By remembering the images or words, you can recall the memory of relaxation in your body and mind and reenter the state of relaxation much more easily and quickly.

GROUND THE MEDITATION

Immediately take the proper steps to ground your experience by bringing it consciously into your beta waves. Write down or speak out loud the keys and landmarks that you noted. You can write your experience in a journal, draw or paint it, speak it into a tape recorder, tell it to a friend, or simply tell yourself, making it conscious in your own mind. These landmarks are the biofeedback that you can use as a guide to help you return to this state of relaxation more quickly and easily next time.

SUBJECTIVE EXPERIENCES—FINDING THE LANDMARKS
"How can I tell where I have been without biofeedback equipment?"

The table entitled "Subjective Landmarks" is designed to give you information about the relationship between a large number of subjective experiences you may encounter during your meditation and the probable brainwave and body states that accompany each experience. At first, simply read through the main column listing subjective experiences. These are the *most common* sensations and perceptions that people have reported during their meditations. They have been collected over years of research with meditators. I have built on both the research by Terry V. Lesh at the University of Oregon on response patterns to meditation and the years of research by C. Maxwell Cade in London with his students. I have then added my own observations from the past twenty years to complete this chart. However, the landmarks in this chart are by no means *all* of the sensations or experiences people have while meditating. Just use them to try to get a general feeling for your subjective landmarks in each category.

SUBJECTIVE LANDMARKS

#	Subjective Landmarks	ESR	EEG
0	May have difficulty stilling the mind *or* Mind racing out of control Itchy, distractible, inattentive state A feeling of "Why am I doing this?" Just beginning to relax A feeling of "settling down"	25 to 20	Continuous beta, often with some flares of other waves Possibly intermittent alpha

1	Foggy state
	Feeling dizzy
	Sensations of going under an anesthetic
	Occasional feeling of nausea
	Mind filled with everyday affairs—almost as an
	  avoidance of inner stillness
	A feeling of scattered energies
	A sensation of drifting off to sleep or being pulled
	  back from the edge of sleep
		20
		to
		16
			Somewhat reduced
			beta, but still
			present
			Intermittent but
			stronger alpha

2	Scattered energies beginning to collect Beginning to feel calmness and relaxation Uninvited vivid flashes of imagery Childhood flashbacks Images from distant to immediate past Attention not very sustained A feeling of being in between states Transitional state	16 to 14	Reduced beta Stronger alpha— could be continuous Intermittent (low- frequency) theta
3	Greater sense of stability Well-defined state Pleasant bodily sensations of floating, lightness, swaying, or rocking Occasional slight rhythmical movement Concentration easier and stronger Increased and clearer imagery Increased ability to follow guided imagery	14 to 11	Highly reduced beta Continuous alpha Possibly more continuous theta with increased frequency and/or amplitude
4	Extremely vivid awareness of breathing Extremely vivid awareness of heartbeat, blood flow, or other bodily sensations Feeling of loss of body boundaries Sensation of numbness in limbs Sensation of being full of air	11 to 8	Highly reduced beta Continuous alpha Increased theta

	Sensation of growing to great size or becoming very small Sensation of great heaviness or lightness Sometimes alternating between external and internal awareness		
5	Very lucid state of consciousness Feeling of deep satisfaction Intense alertness, calmness, and detachment Sensation of spacing out or disappearing from environment and/or body Extremely vivid imagery when desired Feeling of altered state lacking in previous levels, 0 to 4 Sense of peak experience, "ah-ha" experience, intuitive insight High performance	8 to 5	Strong beta mastery —ranging from no thoughts to creative thoughts Continuous alpha Continuous theta
6	New way of feeling Intuitive insight into old problems, as though seen from a more aware level Synthesis of opposites into a higher union Sensation of being surrounded in light A feeling of higher spiritual awareness A sensation that nothing matters other than just being The experience of bliss The experience of indefinable peace A feeling of greater knowledge of the universe	5 to 0	Four Possible patterns: 1. Awakened mind (beta, alpha, theta, delta) 2. Optimum meditation (alpha, theta, delta) 3. Very little electrical brain activity (two straight vertical lines) 4. Evolved mind (circular pattern, including beta, alpha, theta, and delta with no bottlenecks)

To begin with, don't think about what brainwaves relate to which landmarks. Simply peruse this list and notice what you experienced. If you don't understand a phrase, it probably means that you did not experience it. Without trying to figure it out, go on to the next. No matter how brief a sensation might be, make a mental note of it or check it in the book with a pencil.

It is important that you identify even the briefest experience in a category. Identifying it and thereby acknowledging that you had it will help you to return to this state more easily and quickly. Noting your passage, however brief, through a type of experience is like opening a channel to that altered state. Even though the channel may initially be very narrow, it is open. And each time you return to it, your journey will become easier, your stay will be longer, and the channel will be stronger.

If, on the other hand, you downplay an experience that you have, you undermine that experience by denying its validity. If you don't admit to yourself that you had the experience, then the next time it happens, your subconscious tendency will be to deny it rather than welcome it.

WHY SHOULD WE WELCOME THESE EXPERIENCES?

This table can provide a concrete road map for you, on a journey that others have traveled before you, plotting the course. Each time you experience one of these landmarks, you become more grounded on that journey. When you become familiar with this territory, you will develop the ability to go more quickly where you want to go by calling up these subjective experiences within yourself. Remember, you may not experience them in this particular order. You are quite likely to skip around, going back and forth between the different categories. Let's look closely at each category.

CATEGORY 0

You are just settling down to begin the act of relaxation or meditation. People get stuck in this state when they cannot shut off their beta and stop processing the outside world or dealing with their inner anxieties. One of the best ways to move on beyond this state is to become very present in the moment. Let go of your pull back into the past and your hooks into the future and become aware of whatever *is* right now—your breathing, body position, temperature, sensations, and so forth. Deep psychophysical relaxation will also help to bring you through this state, as well as the beta mastery exercises in Chapter Four.

CATEGORY 1

This is a continuation of Category 0, but your hooks into the outside world have loosened and you may be beginning to have a sense of being in an altered state. The tools discussed above—being present in the moment and deep relaxation—will also help you move through this category to a deeper level. A word of caution however—from here down, the path splits into two channels. One takes you into meditation and the other takes you into sleep. This is your first brush with the danger of falling asleep in meditation, and the place where people often do if they are going to. If you feel sluggish or drawn into sleep, take several deep, rapid breaths to arouse yourself slightly. Come back up into Category 0 and start over. Positive affirmation that you are going to remain alert and awake throughout the meditation can also prepare you to drop down the right path instead of dropping into sleep.

CATEGORY 2

Some people skip this state altogether. Some pass through it fairly easily, and others may get stuck here. In Category 2 you experience the first true sense of being in an altered state. This feeling can be confusing because your brainwaves are often not very well organized. You may experience a discordant, slightly jangled sense or a sense of waiting for something to happen, without being sure what that is. You may see hypnogogic imagery—sometimes brilliant flashes of colorful visions—but the visions tend to lack depth and meaning, or they may be related to events in your memory. This state often feels uncomfortable in contrast to the pleasurable sensations experienced in the next category. If you get stuck in this state, deepening your body's relaxation level is by far the best means of breaking through the barrier into the next level down. Often, one major attempt to *let go*, exhaling your breath deeply and relaxing every muscle in your body at once (especially jaws, tongue, and throat area), will carry you through the barrier down into the next level.

CATEGORY 3

Entering this state can be quite a relief if you have been stuck in Category 2 for any length of time. Some will move right here without any problems, while others will knock on the door for some time before they let go and allow themselves to come through. In this category we find our first real state of meditation. Although it is light meditation as opposed to deep meditation, it has all of the biological functions that define the state. Once you have reached Category 3, you will have clearly chosen the path of meditation over the path of sleep. It is possible to meditate in this state for a long period of time and feel satisfied and productive in your meditation. It is a safe place to be and offers physical, mental, emotional, and spiritual rewards. If you are having trouble progressing beyond it and want to go deeper, the major keys are to relax even more and, especially, to work on developing your theta brainwaves. Techniques for this will be explained more thoroughly in Chapter Three.

CATEGORY 4

There is often a barrier between 3 and 4, primarily because Category 4 is another potentially uncomfortable state. It is common to enter it and stay for a brief period of time, only to quickly arouse again to the safety and comfort of Category 3. Once the sensations and experiences of Category 4 are accepted, both as valid meditation experiences and as common to many people, it is possible not only to look forward to them but also to begin to generate them intentionally in order to help you get to this depth more quickly. The subjective experiences in this category usually fall into two classes—extremely vivid awareness of the body, its functions, and its sensations, or complete loss of awareness of the body, accompanied by sensations such as expansion and lightness. People may mistakenly think that, if they are experiencing the first class of symptoms of extreme body awareness, they couldn't possibly be meditating. They believe that they have somehow gotten out of their meditation, so they immediately try to return to Category 3, where they *know* they are meditating. This may result in their having difficulties getting beyond Category 4 into the even deeper and more rewarding meditation experiences of Category 5. It is important to recognize and validate these Category 4 experiences as being signposts and road marks on your deepening meditation journey. Allow yourself to be *fully* immersed in your physical sensations, then let go even more. Only in this way can you move through this state to an even deeper one.

The key to success in Category 4 is to maintain *self*-awareness while losing body awareness. Achieving this will lead you quickly into Category 5. The loss of self-awareness might result, again, in sleep.

CATEGORY 5

When the barrier is broken between 4 and 5, there is a sense of awareness, lucidity, and serenity not experienced in the previous categories. Herein lies the awakened mind. You will note that the brainwave characteristics of this category include beta mastery. This means that you can produce beta when you want to and let go of it when you want to. Therefore, this state is *either* one of deep continuous meditation without beta *or* the awakened mind with beta. The difference between these two states can be found in the content of the meditation. If there is any material that you are processing, developing, creating, or healing, then beta waves may be necessary, and the awakened mind state will result. If, on the other hand, the meditation is one simply of intense alertness, calmness, and detachment, with no content, then the beta is not needed and the brainwave pattern is one of deep meditation.

Many people know when they are in this state because they experience a feeling that is lacking in the lighter meditation states. Once this state is entered, the meditator tends to settle here for the remainder of the meditation, and the fluctuations and vacillations of the previous categories are forgotten. The only things that commonly may arouse you from this state are disturbances from the outside, such as the telephone or doorbell, or disturbances from the inside, such as insights arising from the content on which you are working; these bring your beta back into play. Otherwise, this state has a very stabilizing effect on your meditation process.

CATEGORY 6

Entry into this next category is not nearly as easy as entry into the previous five categories. This does not imply that those are all easy to enter. Usually only long-term meditators experience Category 6 for any length of time. There are several brainwave states that may accompany this experience (see figure):

1. *The awakened mind.* What differentiates the awakened mind in Category 5 from the awakened mind in Category 6 is the depth of experience and the fact that it is less content oriented. There are few, if any, words to describe what an awakened mind pattern feels like in this state. The beta, likely to be of a low amplitude and low frequency, will contain very little *personal* processing. You will experience, instead, a sense of processing information that is *universal* in nature.

2. *Optimum meditation pattern.* In Category 6, this pattern represents the ultimate meditation.

3. *Very little electrical brain activity.* If an individual displays the two straight lines that are indicative of this lack of electrical activity, as well as the experiences described in Category 6, he or she may be having an out-of-body experience. In such a case the brainwaves do not accurately reflect the state of consciousness because somehow the link between the brainwaves and the subjective perception of the state has been diminished or even severed. I have also seen individuals display this pattern with *awareness* during times of extreme fear or unexpected shock.

4. *Evolved mind.* What happens when the yogi or advanced spiritual practitioner who *lives in the awakened mind state* sits down to meditate and go higher? Most spiritual traditions describe a state of

Category 6
POSSIBLE BRAINWAVE PATTERNS

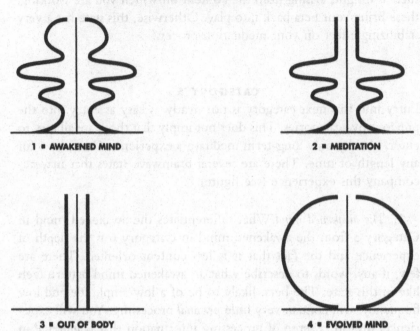

1 ■ AWAKENED MIND   2 ■ MEDITATION

3 ■ OUT OF BODY   4 ■ EVOLVED MIND

the very highest awareness or evolution where the unconscious mind itself has become conscious. In other words, the unconscious mind ceases to exist as a place of separation. There are no longer any bottlenecks or divisions between the conscious, the subconscious, and the unconscious. In this state, one feels a unity with all the universe.

I have not met anyone who lives in this state continuously; however, I have seen a number of people attain it briefly. I have also seen approximations of it, or patterns that appear to me to be stages of the developmental process of this state. One student attained this state for an extended period of time during a meditation on pure consciousness in one of my workshops. She had tears streaming down her face during the meditation and was unable to speak for some time. Afterward, the only word she could say was "bliss" . . .

The list of subjective landmarks presented in this chart is meant only as a guide. Although they are the most frequently reported experiences, they are by no means the *only* subjective experiences available.

Within any particular brainwave category, you do not need to experience all of the sensations on the list to be in the brainwave state that they describe. You may experience only one or two in each category. Several may occur at once, or they may come in succession, one after the other in random order. You may become familiar with all of the sensations or with only a few.

Likewise, with the different brainwave categories shown above, you may experience them in the order I have listed, or you may skip around. Remember, your brainwaves are constantly moving. Though they may stay in one state for a period of time, they may also jump from state to state.

BODY-MIND RELATIONSHIP

So how does the relaxation of our bodies relate to the brainwave states we are producing? The easiest way for most people to begin to consciously control their brainwave states is with a relaxed body, but this does not always occur, nor is it always desired or essential. The major concept that needs to be understood here is how the body and the mind work together to determine your psychophysiological state. This process is illustrated in the Mind-Body Graph (page 44).

In this graph, the horizontal axis represents the body. The far left side represents the body at its most relaxed, and the far right side represents the body at its most aroused. The vertical axis represents the mind—a continuum of consciousness from beta down to delta. For simplicity's sake, we can say that an overly aroused mind is at the top of the vertical axis while a very relaxed mind is at the bottom. Arranged this way, the chart's four quadrants represent four mind-body relationships:

1. aroused mind and aroused body
2. relaxed mind and aroused body
3. relaxed mind and relaxed body
4. aroused mind and relaxed body

Mind-Body Graph

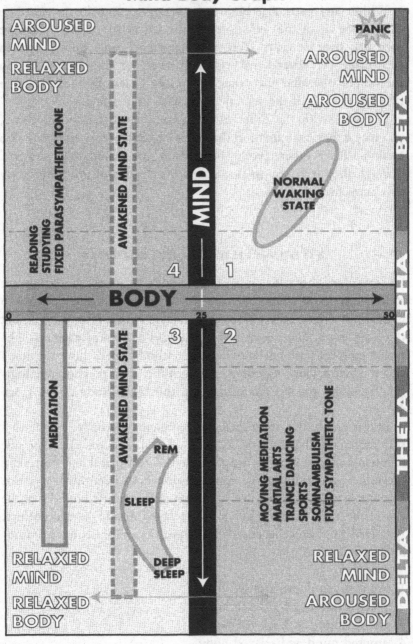

■ 44 ■

There is no place on this graph that is bad or wrong. What is important is to learn *how to be where you want to be, when you want to be there.* The more familiar you become with the experiences that occur in each quadrant of this chart, the greater will be your mastery over your location at any given time.

QUADRANT 1:
AROUSED MIND AND AROUSED BODY

In the top right-hand quadrant, we find our normal (or most typical) waking state. The brainwaves are primarily beta, and the body is somewhat aroused. Taken to its greatest extreme, this state of mind-body arousal results in *panic,* shown in the upper right-hand corner. When the mind and body are overly aroused, your thoughts go crazy, your heart pounds, and your body-mind becomes stressed to the limit. If this experience becomes habitual, stress-related diseases will occur more easily and more frequently. (And how many illnesses aren't in some way stress related?) Most people spend the majority of their time meandering around this quadrant.

QUADRANT 3:
RELAXED MIND AND RELAXED BODY

The next quadrant in which we spend the most time is number 3. Where we typically sleep is marked by a rectangle. The mind moves between the deep-sleep brainwaves of delta and the theta brainwaves of REM (rapid eye movement) dream states on the vertical brain-wave scale. The sleep period for many people ranges between about 13 to 17 on the horizontal body-relaxation scale. This is not nearly as relaxed as the body could be while sleeping. It is possible to go down into the *deep* relaxation that would be found below 5. Meditation that involves deep psychophysiological relaxation is often much more relaxing to the body than the typical sleep states. In fact some people on waking in the morning feel that they have not truly rested all night, that their bodies have been active or tense during their sleep.

Meditation can take place anywhere in the lower two quadrants, where alpha and theta brainwaves are located. However, the more you want your body to receive the stress-reduction benefits of the meditation, the farther to the left in Quadrant 3 you should go.

QUADRANT 2:
RELAXED MIND AND AROUSED BODY

This quadrant contains the same brainwave states as Quadrant 3—some alpha, theta, and delta—the brainwaves of meditation. As we have seen earlier, it is often easier to learn to meditate when the body is relaxed; however, it is by no means essential. In fact many forms of meditation take place while the body is consciously and intentionally aroused. All of the moving meditations (except perhaps the slowest forms of yoga) take place somewhere in this quadrant. Martial arts, such as tai chi, *chi kung* (or *qi gong),* judo, karate, and aikido call on the specific action and arousal in the body to help elicit the meditation state. It is therefore appropriate, indeed essential, that these forms of meditation practices are carried out in this quadrant. Sufi whirling and trance dancing likewise use movement to actually access the altered state.

Sports like skiing, tennis, and bicycling can also move one into an altered state of consciousness. Whether the altered state is the primary goal or not, this by-product can still be enjoyed by the practitioner. Dancing certainly belongs in this quadrant and has long been used to reach altered states of consciousness. Sex also gives us access to altered states. Any and all of these activities can take place in Quadrant 1 as well, with reduction or elimination of the meditative quality.

QUADRANT 4:
AROUSED MIND, RELAXED BODY

We find ourselves most often in Quadrant 4 when we are reading, studying, or performing some other thought-provoking activity that does not involve the arousal of the nervous system. Though a comfortable place to reside temporarily, Quadrant 4's deep physical relaxation can draw the mind down into the sleep of Quadrant 3 at inappropriate times.

Consider the student sitting in the lecture hall who begins the lecture with every intention of staying awake and listening and learning. The body gets very relaxed and the mind soon follows. It may become hard to think or concentrate on the complex information being offered. Several sharp, deep breaths to attain physical arousal may be the most effective way to draw the mind back to a beta state, if that is where you want to be.

Quadrant 4 can also house forms of insomnia. Have you ever lain in bed at night, physically exhausted and wanting to sleep, but with your mind on fire, perhaps worrying or excited? This can be an uncomfortable place to be and may indicate a body-mind split.

BODY-MIND SPLITS

Quadrants 1 and 3 illustrate the body and the mind working together in the same mode. Quadrants 2 and 4 represent what may happen when the body and the mind are moving in different directions. If you want your body and your mind to be operating in the same mode and you find yourself unable to extricate yourself from either Quadrant 2 or Quadrant 4, you may have what is called a *body-mind split*.

Sleepwalking, or somnambulism, is perhaps one of the most extreme examples of this body-mind split. The mind is asleep and the body is active and moving around. Some forms of insomnia also fall into this category—the mind is exhausted and wants to sleep, but the body is tense and unable to let go.

"IS IT POSSIBLE TO BE IN DEEP MEDITATION AND *THINK* THAT MY BODY IS RELAXED, WHEN IT ACTUALLY ISN'T? COULD I BE MISSING THE *PHYSICAL* BENEFITS OF MEDITATION WITHOUT REALIZING IT?"

Some people have a mistaken idea that meditation and relaxation are synonymous in terms of the beneficial effect on the body. Not so!

Without the aid of biofeedback instruments to measure your body's response to meditation, it is quite possible that, although your mind is experiencing deep meditation and you are producing the brain-wave patterns of meditation, your body is remaining tense and unaffected by your inner journey. You may be thinking you are receiving the generally accepted benefits of deep relaxation when, in fact, you aren't—another form of body-mind split.

If you are meditating in Quadrant 2 without an intentional physical aspect to your meditation, you can enter into what is called a *fixed sympathetic tone*. The sympathetic nervous system—that part of the autonomic nervous system that activates with stress reactions—is fixed on *go*. The body is in a state of persistent arousal; however, the mind is still able to meditate. This is why you need to watch for keys to a relaxed body, such as the pleasant body sensations of floating, rocking, and lightness, and the release of muscles.

Fixed sympathetic tone can be experienced in other ways as well. Sometimes a person might awaken from the meditative state by moving from Quadrant 3 to Quadrant 2, instead of going directly to the normal waking state of Quadrant 1. This experience can be strange and frightening without guidance. The body is awake and may even shoot into the high-anxiety range of the fight-or-flight response, and the mind is still in deep meditation, seemingly unable to emerge from the depths. When this occurs, rather than try to drag the mind out of its altered state forcefully, relax the body once again and *return to Quadrant 3*. Then start your arousal process all over again, this time bringing your mind along with your body.

Another form of body-mind split occurs in Quadrant 4, with *fixed parasympathetic tone*. The parasympathetic nervous system is that part of the autonomic nervous system that activates the relaxation response. An extreme example of this split occurs if your mind awakens suddenly or violently from a deep meditation that perhaps released some disturbing content from your subconscious. Your body remains in deep relaxation. The mind is panicking, and the body feels paralyzed, unable to move out of the state of release. Rather than force your body to move from Quadrant 4 up to Quadrant 1, the best action you could take would be to relax your mind again and return to Quadrant 3. When you are ready, start your arousal process once again, this time bringing your body with your mind.

THE JOURNEY FROM NORMAL WAKING STATE TO MEDITATION IN THE BODY AND THE MIND

The meditator's journey is rarely a simple direct line from a normal waking state down to meditation. You might wander throughout the four quadrants, sometimes meandering around in one, or sometimes jumping abruptly from one to another. If your meditation deepens, you may end in Quadrant 3 and rest there for some time before reawakening. The graph on page 50 illustrates a potential journey into meditation. Initially, the meditator relaxes the body and mind together a little bit. Then the mind takes a dive into deeper alpha and theta states. The body starts to catch up, but the mind arouses a little. From here, the body and the mind again relax together, and a pleasant state of meditation is attained for a very brief period. The meditator may then become self-conscious, thinking, "Oh my goodness! I'm meditating!" and shoots out of the meditating state back up to the waking state to begin the journey again.

Each time you go down into a meditative state, your experience will be a little different. With practice, the meditator finds the journey more under conscious control and is able to move more directly from Quadrant 1 to Quadrant 3. As you gain self-mastery, you will be able to go anywhere you want on this chart and recognize the physical and mental sensations that accompany being there.

When an awakened mind state is attained, it encompasses the entire vertical range of brainwaves from beta down to delta. With enough mastery, you can maintain this state of consciousness, whether your body is in deep relaxation, high arousal, or moving somewhere in between (see page 44).

Now look back at the table of Subjective Landmarks and read across all four columns. *If your body and your mind are relaxing together,* the first column indicates the number of the category, the second column gives the subjective experience you may be having, the third column gives the relaxation level likely for that subjective experience, and the fourth column names the brainwaves that most often correlate.

Mind-Body Journey

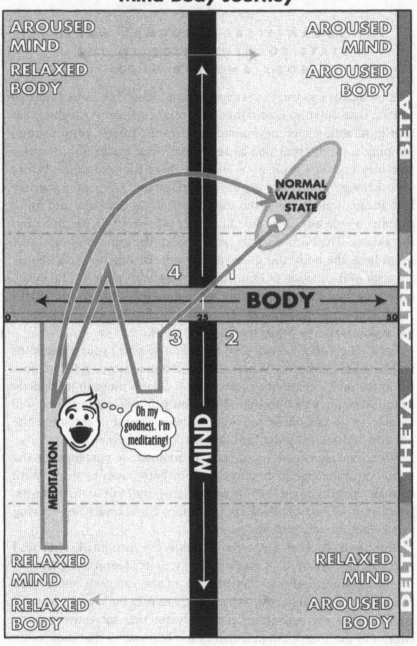

Hand temperature can also be a very helpful source of body-mind information. When you relax deeply, your hands generally become warmer. When you are aroused, they become cooler. There are exceptions to every rule, however. Using hand-temperature biofeedback, I have known people to have extremely warm hands while not experiencing a state of relaxation or meditation at all. I have also seen people have an awakened mind state, new insights, or peak experiences while retaining very cold hands.

Practicing mastery of arousal and, especially, of relaxation of your nervous system is a major step toward the mastery of your brainwaves and your internal states of consciousness. If you begin your meditation with conscious relaxation, you will have more assurance that the body will also get the physiological benefits that relaxation brings. You will find relaxation helpful in developing the different categories of brainwaves, regardless of whether you have excess in some and deficiency in others. We need to find a neutral ground from which the brainwaves may grow. With this in mind, we will move into the development of the brainwaves of meditation.

III

MEDITATION
AND HIGHER STATES

Meditation is perhaps the most commonly talked about and accepted altered state of consciousness, yet there are many different definitions, methods, philosophies, dogmas, and techniques. Although attitudes about, approaches to, and uses for the mental states called meditation vary with the popular trend of the time, a recurrent and consistent theme runs underneath all of the changing variables of meditation. The theme is that meditation, done properly, provides a *higher* state of consciousness than our normal waking states.

WHAT DOES ''HIGHER'' MEAN?

To many the word "higher" (as in "higher states of consciousness") means more spiritual, more attuned to some form of God-consciousness or to a higher power. To others it means greater creativity, productivity, and the ability to access imaginative, original, inventive, and ingenious ideas. And to still others it means increased mental and physical health and well-being, accessing powers of self-healing or personal growth.

''SO *HOW* CAN WE DEFINE MEDITATION WITH ALL OF THESE DIFFERING OPINIONS?''

When discussing meditation, the most useful way for us to find a common language is to talk about it in terms of the brainwaves that people produce when they are meditating. Interestingly, despite the differing descriptions, spiritual backgrounds, religious beliefs, purposes, and methods of meditation, there is a common combination of brainwave frequencies that underlies *all* of them. With slight variations, the brainwaves of meditation are alpha and theta—with or without delta, depending on the style of meditation.

Figure 1A shows that the brainwaves of meditation are symmetrical in both hemispheres and that the alpha is stronger in amplitude than the theta. The theta provides the depth of meditation—the subconscious inner space from which the creativity springs, or the spiritual connection is made, or the self-healing is programmed in the body. Theta gives you the experience of profundity in your meditation. Alpha provides the link, or bridge, to the conscious mind so that you can actually remember the contents of the theta. If you meditate without alpha, producing theta alone, you *will not remember your meditation!* Remember, theta is the *sub*conscious, and theta waves generally occur *below* conscious awareness. So anything that takes place below that line will remain *sub*conscious unless the brainwaves that bridge the subconscious to the conscious are active. When acting as a bridge, alpha waves provide a clearer, sharper form of imagery through which the contents of the theta can be filtered.

There are many different words for types of trance states. Meditation, contemplation, concentration, prayer, hypnosis, guided fantasy, visualization, and deep relaxation are the ones most often used. The brainwave patterns are the same or very similar for all of these variations of trance state. The difference is found primarily in the method of induction and the goal while in the state, rather than in the state itself.

There are exceptions to this rule, however. Sometimes the purpose of the meditation is active, requiring analysis or thought. Then beta is added to this pattern. For example, problem-solving meditations often ask that the meditators enter a deeply relaxed state of meditation, then use particular imagery techniques to access solutions to predetermined issues or problems. This form of meditation usually requires that the meditator add beta brainwaves to the already existing meditation state of alpha and theta. In this case the brainwave pattern can be surprisingly close to the awakened mind pattern. This illustrates why certain forms of active meditation are used in

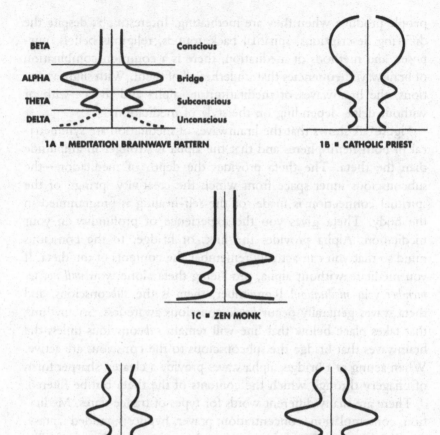

BETA

ALPHA

THETA

DELTA

Conscious

Bridge

Subconscious

Unconscious

1A ■ MEDITATION BRAINWAVE PATTERN

1B ■ CATHOLIC PRIEST

1C ■ ZEN MONK

2A ■ ZEN MASTER

2B ■ WITH RELAXED TONGUE

awakened mind brainwave training to add the beta to the brainwave pattern once a stable alpha and theta meditation pattern has been established.

There are strong similarities between the brainwave patterns of meditation and prayer, but there are also some differences. For example, there is a marked difference between the pattern of a Catholic priest in a state of prayer and a Zen monk in a state of meditation.

I measured the brainwaves of a Catholic priest from the East Coast who was interested in finding out how he could make his prayer deeper and more spiritual. He was going on a residential retreat for a month and wanted my input so that he could make the most effective use of the time he was away. In looking at his brainwaves, I saw that he did indeed have an awakened mind brainwave pattern. But as you can see from Figure 1B, his brainwaves are *very* top-heavy! He asked for my advice about how to pray, and when I suggested "Fewer words!" he laughed and said he knew exactly what I meant. This individual really felt his spiritual connection with active theta brainwaves and maintained awareness during his prayer with open alpha waves. However, the bulk of his prayer was conducted by subvocalizing words—hence his large beta.

On the other hand, a Zen monk practicing simple sitting with no thought might have the brainwave pattern in Figure 1C. Here there is no beta at all—no words or thoughts in this person's head! But you can see that the delta is very active. This person has added unconscious awareness to the meditation. The practice of Zen is essentially a nonverbal experience, awareness of whatever IS being the main tenet. Hence the Zen practitioner is likely to have less beta and potentially more delta than a person in active verbal prayer.

One longtime Zen master whom I monitored had an awakened mind pattern (Figure 2A) with the exception of the high-frequency beta flares. The beta rounded out (Figure 2B) as soon as I told him to relax his tongue. This was an extraordinary idea for him, and he said that he "couldn't wait to tell my students about this," that it would revolutionize his way of teaching.

"IS FOLLOWING A GUIDED FANTASY REAL MEDITATION?"

If the guided fantasy makes you produce very little beta waves and strong alpha and theta waves, the answer is a resounding *yes*. If your response to the guided fantasy is to think and analyze instead of allowing the images to simply be present without consciously creating or changing them, then the answer is no. In fact, guided fantasy is a very good entry into the experience of meditation for most people. Some traditional meditators may argue that guided fantasies fill the mind and that the purpose of meditation is to empty the mind—a good argument from the perspective of *content*. Not so from the perspective of *state*. The state of meditation is one of alpha and theta brainwaves. Anytime you are producing those brainwaves in combination without producing many beta waves, you are in a state of meditation. Guided fantasy can lead you to these alpha and theta brainwaves potentially much more quickly than sitting and watching your thoughts until, hopefully, they disappear.

"HOW IS GUIDED FANTASY BEST USED AS A MEDITATION?"

Guided fantasy for the purpose of meditation should have several key elements.

First, there should be some form of relaxation to allow the listener's body and mind to begin to relax. Slowing the breathing down and reducing the stress levels in the body help the listener to begin to alter his or her brainwaves from the normal waking beta state to that of meditation.

Second, imagery helps the listener produce alpha waves. Hence the fantasy actually begins to help produce the waves that will bridge you into the meditation state.

Third, the guided fantasy should have suggestions that will take you deeper into yourself, into the theta waves of the subconscious.

And fourth, once you are there, the guided fantasy should give you ample time in *silence* to experience the state of meditation without external input.

If the guided meditation follows these principles, then the listener may be able to enter a state of meditation even during the first few attempts. This doesn't mean that he or she will be able to stay there, however. The way that the meditation ability develops in even a consistent practitioner is generally in brief flares or flashes. The brainwaves are constantly moving, shifting, and changing—sometimes radically, sometimes gently. Often during the first attempts at meditating, no matter which entry method is used, the practitioner passes through a state of meditation several times. Any number of things may happen to shift the pattern quickly out of a state of meditation soon after it is attained. The most common distractions are beta waves, thoughts that interrupt the inner stillness. Another problem is losing the alpha waves, so that you are still awake but not conscious. Still another problem is falling asleep.

The key to successful meditation is to recognize and validate those periods of time when you are meditating. You may have an urge to denigrate that moment of clarity or insight as something that you "just imagined." Or you may tell yourself that it didn't count because it went away so quickly. Don't do that to yourself. Validate every insight, every flare of subconscious imagery that rises to consciousness, every moment of internal stillness and peace. Know that when this happens, no matter how briefly, you are meditating. If you do this, you will find that those moments of clarity and meditation begin to come more frequently and stretch for longer and longer periods.

THE *TECHNIQUE* IS *NOT* THE MEDITATION.

You may hear someone say, "I have to go meditate"—when what they mean is they have to go practice a particular breathing system they have learned. Many people think that they are meditating because of some particular technique they are using or a set of instructions they follow. Rather, they may be simply sitting and thinking without any entry into the meditative state whatsoever. They may wonder why they don't get anything out of meditation. Remember, practicing a certain technique that is supposed to make you meditate does not necessarily mean that you are meditating.

A man came to see me once saying that he had meditated for *an hour a day every day for twelve years.* Although he enjoyed the time he spent sitting, he felt he was missing something. From talking to other meditators, he felt that he must be doing something wrong because he had none of the experiences that he had heard others describe. I measured his brainwaves while he was "meditating" and discovered that *he had spent those twelve years simply thinking!* The alpha-theta state of meditation was actually unfamiliar to him. Within three sessions I guided him into the experience of deep meditation. When he first reached it, he sat bolt upright with a big grin on his face and cried, "I got it!"

Granted, his case was a rare extreme; but various manifestations of this misconception about meditation are more common than most people would care to think. There is an inherent danger in getting locked into the technique instead of the meditation.

Many people carry some form of secret apprehension that they may not be doing it right or may not even be doing it at all. Because they really want to meditate, they *try even harder,* practicing their technique with even more determination and zeal. This can begin a vicious cycle. The more they try, the less they seem to be able to meditate. You cannot *try* to meditate. Meditation is a process of *allowing,* not trying.

You can tell in several ways if you are meditating. Signposts exist along the paths of entry into meditation that will help you understand what brainwaves you are producing and where you are in relationship to the alpha-theta brainwave state you are seeking. By studying the chart on page 35 (in Chapter Two), you can begin to understand what brainwaves your experiences may be reflecting. Your journey down into meditation might involve some or all of the those experiences, but not necessarily in that particular order. Examples of brainwave states as you journey into meditation may include the following:

CONTINUOUS BETA WITH POSSIBLE FLARES OF OTHER FREQUENCIES

This pattern often occurs at the very beginning of a meditation or following a disturbance during the meditation. You might be experiencing sensations from Category 0.

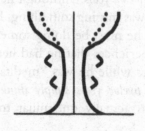

REDUCED BETA WITH INTERMITTENT
BUT STRONGER ALPHA

Here you may be beginning your descent into meditation. The experiences are often those of Category 1.

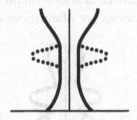

REDUCED BETA, STRONGER ALPHA
(COULD BE CONTINUOUS), INTERMITTENT
(LOW-FREQUENCY) THETA

This is often experienced as a transition into meditation, with sensations predominantly from Category 2.

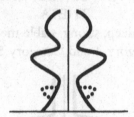

HIGHLY REDUCED BETA, CONTINUOUS ALPHA,
POSSIBLY MORE CONTINUOUS THETA, WITH
INCREASED FREQUENCY AND/OR AMPLITUDE

These brainwaves show a light but relatively stable meditation, such as those experienced in Category 3.

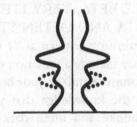

HIGHLY REDUCED BETA, CONTINUOUS ALPHA, INCREASED THETA

These brainwaves show continuing and strengthening meditation and may also be experienced as another transition phase into an even deeper state. The sensations are often those of Category 4.

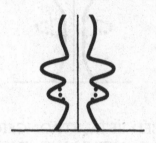

STRONG BETA MASTERY, CONTINUOUS ALPHA (USUALLY LOW-FREQUENCY) AND CONTINUOUS THETA

This pattern is one of deep, strong, stable meditation. Sensations can occur from both Category 4 and Category 5.

VERY LITTLE BETA, VERY LITTLE ALPHA, STRONG THETA AND OFTEN STRONG DELTA

We can subtitle this brainwave pattern as "I was gone, but I wasn't asleep." Experiences may include a sensation of spacing out or disappearing from the environment and/or your body, very little memory of the experience but the knowledge that you were aware at the time. Images are dreamlike and often dark and hazy, sometimes tinged with blues and purples. You may have the sensation that something important has happened or is happening, but you don't know what it is. You don't know where you were, but you *know* you weren't asleep; and when you return, it's as if from a very long way away.

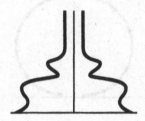

DELTA

You probably have no memory of the experience at all; and when you return, it is quickly, with a startle or a jerk. You were asleep!

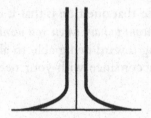

BETA, ALPHA, THETA, AND DELTA

Here we have the awakened mind. In this state you may experience all of the sensations of meditation, with the added ability to think. You may have any of the experiences from Category 5, especially a strong intuitive insight.

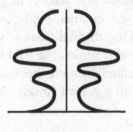

BETA, ALPHA, THETA, AND DELTA WITH FEW OR NO BOTTLENECKS

This is a transcendent state and is often accompanied by experiences from Category 6.

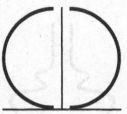

WHAT IS THE BEST PATTERN TO HAVE?

This series of brainwave patterns should not be viewed as a hierarchy. Who is to say that having theta is more important or better than having beta? Even having an awakened mind might not be something you truly want to experience *all* of the time. So, what is best? The only real answer to that question is that it is best to *be able to have the brainwaves that you want to have when you want to have them*. In other words, we are working toward being able to alter your state of consciousness at will in accordance with your needs at the time.

"I DON'T MEDITATE, BUT I HAVE SOME OF THE EXPERIENCES DESCRIBED AS MEDITATION!"

Remember the fellow who thought he was meditating but found that he was thinking instead? The exact opposite of this can also take place. A young woman came to me supposedly for relaxation training and stress management. She was, in fact, suffering from the results of early childhood trauma, and we ended up spending considerable time working through that as we developed her theta brainwaves. During the first session she told me she had never meditated. She was even unfamiliar with the word. I hooked her up to the EEG and took her on a guided fantasy. Almost immediately, her beta reduced dramatically, and her alpha, theta, and delta came out in a beautiful meditation pattern. After the relaxation was over, I asked her if this was a new experience. She said, "Oh no! I feel this way all of the time when I sit in a field of flowers."

Some people who know nothing about the concept and definition of meditation spend time frequently, even daily, in a state of deep meditation without knowing what it is called (and without caring). They may think they are just daydreaming, resting, sitting, listening to music, or—on the active side—tending the garden, walking to work, running, arranging flowers, cleaning the house, or painting.

In terms of your brainwaves, the combination of alpha and theta without beta is meditation, regardless of which activity you are engaging in to get you there. For the purposes of this chapter, I am defining meditation as *any still form of contemplation and relaxation that produces those brainwaves.*

A THETA-ACCESSING MEDITATION

Before we go any further with the theory of meditation and brainwave training, I would like to give you a meditation designed to access theta brainwaves. These are the brainwaves that are usually the most lacking in people and consequently the ones that need the most immediate work. Practice this meditation first before reading on. The discussion that follows will explain *why* we are doing what we are doing. I've always found it more effective to have my clients or workshop participants try a meditation first, and then learn the theory about it afterward. This way their minds are not cluttered with theory when they are first doing it.

Have fun!

Please note that * indicates *pause:*

(*)   = 2 to 5 seconds
(**) = 5 to 10 seconds
(***) = 10 to 30 seconds

I will note longer times. Use this as a period of silent time to follow the instructions and to meditate. Initially, stay in the place of stillness and silence only as long as you are comfortable. If your mind starts to wander, or you become bored, or you are finished with your explorations, move on to the next phase.

HOUSE OF DOORS

Let your mind clear, and focus on your breathing.
Spend at least five to ten minutes relaxing before continuing. (You may
   wish to refer back to the relaxation beginning on page 31 as preparation for this meditation.)

And now, from this place of deep relaxation and peace, (*)
in your mind, create an environment. (*)
Imagine or create an environment outdoors, somewhere pleasant for
   you. (*)
It could be the mountains, the country, the beach . . . (*)
See it clearly. (**)
Walk around it in your mind. (*)
Using as many senses as possible, explore this environment carefully.
Experience the colors, the shapes, the textures, the sounds, the smells.
   (**)
What is the time of day, the atmosphere, the temperature?
Feel them with your mind. (**)
Make any changes that you would like in this environment to make it safe
   and secure for you.

(pause for one or two minutes)

Now, within this environment, place a house.
Don't go inside yet; just see it from the outside. (*)
We're going to go on a journey through this house. (*)
So, going in through the front door, (*)
and passing through an entrance hallway,
you find yourself in a room covered in mirrors. (*)

Mirrors all around you.
You notice your image, your reflection, in these mirrors. (***)

Now, passing through the room of mirrors,
you find yourself in a long hallway, (*)
a seemingly endless corridor with rows of doors on either side. (*)
And each of the doors is a different color. (*)
You walk down this hallway, passing door . . . after door . . . after
    door, (**)
until you come to stop in front of a door on your left.
You notice the color of the door.
And on the door, there may be a label, an image, or a symbol.
If so, you see it and understand. (**)
When you are ready, you open this door and go into the room that is
    behind it. (*)
Take all the time that you need to explore this room and its contents very
    thoroughly.

(pause for one minute)

While you are in this room, you have the power to make any changes
    that you want. Take the time now to make any changes you wish to
    make.

(pause for one minute)

In a few moments you are going to leave this room. (*)

So allow yourself to take the time now to complete anything that you are
    doing here. (***)
You can always return at any time if you would like to spend more time
    here. (***)

And now, when you are ready, leave this room behind and go back out
    into the corridor. (**)
Close the door behind you, but don't lock it, so that you can return if you
    wish. (**)

And continue your way on down the hallway, again passing door after door, (***)
until you come to stop in front of a door on your right. (*)
Again, notice the color of the door . . . and the label or symbol, if there is one. (**)
When you are ready, open this door and explore what is behind it. (**)
Take as long as you need to very thoroughly explore this room.

(pause for one minute)

Remember that you have the power to make any changes that you want to make at all, within this room.

(pause for one minute)

Now when you are ready, begin to complete what you are doing in this room. (***)
Remember that you can always return here if you wish. (***)

And now leave this room behind. (**)
Go back out into the corridor, close the door behind you, but don't lock it. (**)
And this time, retrace your steps back down the hallway the way you came. (*)
Back past the first door that you entered. (**)
Back through the room of mirrors. (*)
Notice any changes in your reflection, your image. (**)
Back through the entrance hallway. (*)
Back out the front door. (*)
Back into your environment. (**)
Making any changes that you want to make within your environment, find a comfortable place to sit . . . and to meditate. (**)
Meditate on what you experienced in the house of doors (*)
and what it might mean to you.

(pause for one to two minutes)

When you are ready, (*)
find a few "keys" to bring back with you when you reawaken, (*)
to remind you of where you have been, what you have experienced, and
what you have gained from that experience. (*)
These keys may be images, symbols, body sensations, colors, emotions,
words, or phrases . . . anything that will help you remember your
experience . . . even tastes, smells, sounds, and textures that were
present in your meditation.

(***)

In your own time, when you are ready, begin to allow yourself to find a
closure for your meditation. (***)
In your own time, when you are ready, begin to allow yourself to return,
(**)
back to the outside space. (**)
Allow yourself to reawaken and return, (*)
feeling alert . . . and refreshed. (***)

Take several deep, rapid breaths, (**)
and allow yourself to stretch, beginning with fingers and toes. (**)
Open your eyes and reenter the outside space.

GROUNDING THE EXPERIENCE

When you do this meditation you might want to have a pencil and
paper handy. You have probably had a number of experiences within
this meditation that will (a) give you information about yourself and
(b) help you remember what the meditation state of consciousness
felt like so that you can re-create these experiences and return to this
state more easily next time. You may also have some unfinished
business in the house that you would like to go back and finish in a
subsequent meditation.

Because it is the nature of meditation brainwaves to be near or
below the conscious border, meditators—especially those with less
experience—tend to lose or forget the content of the exercise very

easily. Remembering details of the experience will help build the ability to meditate in the future, reinforcing this state as a familiar brainwave pattern.

So how do we bring back content? If you were producing a true meditation pattern of theta and alpha, you will have already created the bridge that allowed you to be aware of the content of the experience while it was happening. If your alpha is low in frequency, however, the material will travel only halfway along that bridge. Because nothing is drawing it strongly enough toward consciousness, it will literally turn around and slide back down into the subconscious (Figure 3A). In these cases, you *know* that you had the experience, and it even feels as if it is there on the tip of your tongue, but you can't quite get it to emerge from the subconscious.

What you need is a tool to bring your experience into full conscious awareness immediately after you emerge from the meditation state. To bring it all the way up into your beta waves, you must put it into thought form (Figure 3B). The best way to do this is through words. Writing or speaking about the experience immediately after you have completed it will draw the contents up into your beta mind so that you can retain them consciously.

If you are practicing this meditation alone, you may put it into words by writing it down. Write down your "keys" first. Then go back and fill in the details. Recount as many particulars as possible. The more minute observations you can remember, the more the meditation will stay with you. Or you may speak it out loud, even into a tape recorder, so that you have a record of it.

If you are practicing these meditations with a friend or in a group, you can describe the experiences to one another. Don't analyze or discuss the meaning at any length until everyone has had a chance to make their content fully conscious. If you talk (using your beta waves) about someone else's experience before you report on yours, you are also likely to lose your content, because analyzing their story may break the link of your conscious mind to your subconscious mind. Sometimes that link is so tenuous that the effort of activating your beta waves for discussion will pull you out of your alpha and theta, and the details of your own experience may fade. After everyone has had a chance to share at least his or her keys, then you can discuss freely and interact without fear of losing your experience.

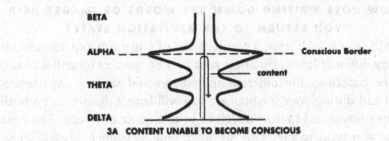

BETA

ALPHA — — — — — — — — — — — — Conscious Border

content

THETA

DELTA

3A CONTENT UNABLE TO BECOME CONSCIOUS

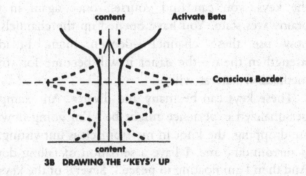

content   Activate Beta

— — — — — — — — — — — — Conscious Border

content

3B DRAWING THE "KEYS" UP

3C RIDING THE "KEYS" DOWN

HOW DOES WRITING DOWN KEY WORDS OR IMAGES HELP
YOU RETURN TO THE MEDITATION STATE?

Meditation is a *felt sense*, a combination of sensations experienced on many different levels. By using your beta brainwaves to verbalize and make conscious the different ingredients and sensory experiences you had during your meditation, you will have a greater access both to its content and to how it feels to be in a meditative state. The next time you want to meditate, sit down and remember this felt sense. By making use of the landmarks, or keys, that drew up the contents of your subconscious mind to your conscious mind, you can simply reverse the route and ride those keys back down (Figure 3C). Using the keys, you can find yourself once again in the meditation-brainwaves state. You have opened up the channels. The more you now use these channels—deepen them, broaden them, and strengthen them—the easier it will become for you to reenter the meditation state at will.

These keys can be many and diverse. An example of someone's personalized experience might be "I'm going down, my shoulders are dropping, the knot in my stomach is untwisting. A soft darkness is surrounding me. I have a sensation of falling down a deep well. And then I am floating in peace." Several of the keys here have to do with descent. The words this meditator might bring back are "down," "untwisting," "soft," "dark," "falling," "deep," "well," "floating," and "peace." Choosing the three most personally evocative is often more effective than trying to remember them all equally. (Once "dark" is remembered, "soft" may well immediately follow anyway.) So this meditator's keys might end up being "down," "dark," and "floating."

Another example is "I am *flying* on the back of an immense *eagle, rising* toward the *light*. I hear a kind of *heavenly music* all around me and feel the *wind* blowing through my hair." The key words are italicized.

Or, briefly, "I feel myself *sinking* to the bottom of *deep*, *dark*, beautiful *pool*." Or "I feel myself *opening*, like a *flower* in full bloom." Or "I see the *sun* rising high above me and feel the *warmth* on my skin." Or "I am surrounded by a *vast tranquil nothing*." It could simply be "the color *blue*," "a sensation of *emptiness*," "a feeling of *expansion*," "the sound of the *waves*."

Although each person's experience is unique to him or her, many common, shared experiences are almost archetypal in nature. Finding your own personal keys, generated inside the experience, is usually much more meaningful than having external keys imposed upon you. However, if you are having trouble getting deeply into a felt sense of meditation, using any of the above phrases might help you.

THE STEP-BY-STEP PROCESS OF BRAINWAVE TRAINING

So far you have looked at the body's relationship to meditation and the importance of relaxation. You have also experienced theta development through the House of Doors meditation. In subsequent chapters, you will continue your brainwave training by learning how to reduce beta waves, access alpha waves, access theta waves, and stabilize delta waves. You will also learn how to develop the appropriate amount of beta waves to experience an awakened mind pattern and develop a high-performance mind. Although you can link into this process at any level, I have found that a step-by-step development can be the most useful for most effectively mastering your brainwave pattern. That is why the exercises follow a particular order in this book. Once you understand each technique, what it is used for, and how it affects you, you may find that changing the order to suit your particular needs will be useful. Through years of practice, I have found that diving into theta development is helpful in the earlier stages of brainwave training.

ACCESSING THETA

The key to accessing theta is to find a way to go deeply into yourself —into your own subconscious. The types of images that best access theta brainwaves are those that take you on long journeys. The actual image itself doesn't matter; it is the effect the image has on your brainwaves that is important. Images that take you *down, through,* and *in* help to access theta brainwaves. The more *changes* you make, the deeper you tend to go. Remember the experience in the House of Doors meditation above. After stimulating alpha waves with the sensual imagery in the outdoor environment, you went *in through* the front door, *passing through* an entrance hallway, *into* and *through* a room of mirrors, *down* a long hallway, and *into* a room. Then the process deepened your brainwaves even more by repeating it on the opposite side of the hall.

To help deepen theta waves, you can also use images that take you *under, over, around,* and *up.* You can create your own theta-accessing meditation by developing imagery using these directional movements.

For example, allow yourself to go . . .

—*down* stairs
—*down* and *around* a long spiral staircase
—*under* an archway
—*into* a tunnel and *down deep into* the ground
—*along* a narrow passageway
—*through* a big wrought-iron gate
—*up* a steep incline
—*inside* a large square structure
—*over* a pile of bricks that is blocking a hidden sanctuary
—*into* the exact center of an octagonal space

Once you have arrived at the end of your journey, you need to have something to meditate about. It is good to set an expectation that you will experience or find something valuable once you arrive in that state.

This expectation accomplishes two things: it gives you a structure around which to meditate, something on which to lean, and it offers you an opportunity to experience content in your subconscious, thereby building *more* theta waves. If you just arrive at the end of your journey and do nothing further, you might still continue to produce some theta, but you might also stop producing theta as soon as you stop your journey. Having some sort of content at the end of your journey gives your theta waves something on which to focus. This content in turn stimulates more theta waves, allowing you to continue to learn something about your own inner workings. This process of self-discovery cannot take place unless you continue to produce theta waves in proper balance with your other brainwaves.

WHAT KIND OF CONTENT SHOULD I USE?

The content can be specific to a particular situation or question in your life, or it can be general and simply focused on a broad category. Examples of specific contents might be:

—which of the two job offers you should accept
—how to solve the design problem on the drawing board at work
—how to handle a delicate communication problem with your spouse
—where the money is going to come from for the Christmas presents
—how to pass your final exam
—what to write for the magazine article

The possibilities are endless. If you are seeking the answer to a specific question, however, you must guard against using beta to solve your problem. When you are trying to develop an awakened mind pattern, you need beta. When you are trying just to meditate, you don't want to engage your thoughts.

So how do you meditate on a specific topic without activating your beta brainwaves? You set up your meditation in advance. Sit down, close your eyes, and get ready to begin, but before you still your mind and empty it of beta waves, think about what you would like to focus your meditation on. If you have a specific question or problem like those listed above, set an intention that you will get some information or guidance during your meditation. *Then forget it!* Let it go completely and begin your meditation.

Later, when you complete your journey down and get to the

place of stillness and meditation, let the information you are seeking well up from your subconscious theta waves. Don't analyze, think about, or even try to understand what you are experiencing. You can save that for later, when you ground the meditation back in your beta waves.

Answers can come in many different ways. One client, an airline flight attendant, wanted to know how to reduce stress in her life. She asked that question of herself and then forgot about it completely. She was enjoying meditating, surrounded by a lovely image of her serene and peaceful inner meditation room, when suddenly the gigantic nose of a 747 airplane burst through one wall and shouted, "Take a vacation!"

Another client asked a very complex managerial question about his company. His meditation imagery was set up outside in the country. A gigantic eagle swept down, landed next to him, and asked him to climb on its back. The eagle then flew my very surprised but enthusiastic client to a small island, where he saw an image of the problem being solved. If he had refused to get on the eagle's back or had thought, "This has nothing to do with my question," and eliminated the imagery, he would have lost both the beautiful journey and the solution to his problem.

Images are not always this distinct and sensory. The answer may come through just a sense of knowing what to do. I have been amused several times by people (myself included) who have confidently stated after a meditation, "I really didn't need to meditate on that problem. I knew the answer all the time!" Of course we did, but we didn't know we knew it before the meditation. We needed the meditation to unlock that knowledge from our subconscious minds so we could be aware of it.

NONSPECIFIC CONTENT

You are at the end of your journey down into the meditation state. You are actively producing alpha and theta brainwaves and very little beta waves. You want to stay in this state and meditate. What do you do to allow this to happen without *working* on a specific issue?

If you have no focus, only the assignment of meditation, it is possible that you will have a deep, meaningful, spiritual meditation experience. It is also possible that you will have a relatively content-free period of quiet stillness. It is further possible that your mind will wander. The squirming of the worm in the brain will return. You will get bored. You will fall asleep. Or you will start thinking—perhaps planning what you are going to do after your meditation.

Try it and see what happens. If you have trouble holding the meditation state, then it is probable that having *some* content on which to focus will be helpful to you. It will simply provide something for you to lean on, an anchor to hold you in the state of consciousness you have achieved through your journey down. If this is the case, give yourself a general topic on which to focus.

You might want to choose from the following:

—connecting with your higher power
—connecting with your higher self
—self-healing
—healing others
—expansion
—withdrawal
—energy (see Appendix on kundalini)
—love
—light

There are many possibilities. *Try out several and see what works for you.* If a certain focus of content helps hold you in the meditation state better than others, use it until you are comfortable letting go completely. You may always find that having a focal point like this to your meditation will give you a greater depth of experience and add meaning to something that might otherwise become just a physiological exercise in stress management.

COMPLETING THE MEDITATION

Don't just suddenly say, "Oh, I've got to stop!" and jump up out of the meditation. This is a recipe for memory loss, as well as an uncomfortable way to conclude. You can use the guidelines I gave at the end of the House of Doors meditation for grounding *any* meditation.

First, before you leave the meditation state, find a few keys to help you remember where you are and what you have been experiencing. Crystalize your experience into a few images, symbols, colors, sensations, or words. As you reawaken, bring those keys into your beta mind by writing them down, speaking them, or at least thinking about them. Reactivate your body slowly. You want your re-emergence to be pleasurable and complete. Start with the periphery, moving, stretching, and wiggling your fingers and toes, and let that turn slowly into a full-body stretch. Take several deep and somewhat rapid breaths.

Notice what it feels like as you reawaken! This will give you important biological feedback that will show you the difference between the meditation state and the waking state. *The more you can become aware of the difference in these two states and the sensations of the journey between them, the more you can learn to master that journey.*

''BUT I ALREADY KNOW HOW TO MEDITATE!''

If you have a method of meditation that works for you, by all means use it! Refer to the descriptions of subjective experiences in this book to get an understanding of what brainwaves you are producing while you are meditating. These descriptions relate to the *state of meditation,* not the method of entry into that state or the content of the meditation.

If you wonder whether another method of entry might give you a different perspective or slightly alter your state, perhaps you can try it without giving up your normal practice. Alternating techniques works for some people. Perhaps just adding a new method occasionally will give you some additional insight and inspiration.

No matter which approach to meditation you use, which philosophy or theology you follow, or whether you are a beginner or a long-term meditator, there are certain common problems that you might, on occasion, encounter. Solutions to these difficulties can often be found by understanding the brainwave activity behind them.

SOLUTIONS TO COMMON MEDITATION ''PROBLEMS''

If you are meditating and still have a hook into the outside world, what can you do about it?

You know you are meditating, but at the same time there is a part of you that is still involved, inappropriately and against your wishes, in the outside world. This can take the form of pure thought, such as list making in your head; mental conversations; or thinking about something you ought to be doing, want to be doing, don't want to be doing, are afraid of doing, or are excited about doing. In other words, you are thinking on top of your meditation.

Sometimes, when you successfully manage to turn off the projections into the past and the future, this problem can still sneak back in by turning the thought onto the meditation itself. In this case you might be thinking about how you are doing, *thinking about the fact that you are not thinking* about the past or the future, or worrying about the fact that you are *going to think.* Beta waves can get into marvelous twists and contortions to stay active!

In all of these situations, your higher-frequency beta waves are the culprit. If you find that no matter what you do with the content, some form of beta pops back in, begin to pay attention to your tongue. It could end up being the most important key to getting rid of those thoughts.

Every time an unwanted thought comes through, breathe relaxation into your mind, and consciously relax your tongue as you exhale. Exaggerate it until you truly get the experience of relaxation into your tongue. At the same time that you are breathing the tension out of your tongue as you exhale, breathe any thoughts away too.

There are subtler ways than thoughts to keep a hook into the outside world. Some low-frequency, high-amplitude beta waves are actually experienced below thought level, almost as if a mental static were going on, or a buzz that won't let you enter fully into your meditation state. You may also experience these waves as an un-wanted awareness of the outside world, something that makes it impossible for you to quite let go of your perception of the room around you.

Because these experiences happen below the common frequencies of verbalization and subvocalization, relaxing the tongue may not easily solve this beta problem. The best solution I have found with my clients is *very deep relaxation*. Going into a deep theta state helps eliminate these holdout beta waves. It is also important to ensure that your meditation space is safe and that you won't be interrupted. If we are unsure of whether our meditation might be disturbed or violated in some way, our innate survival mechanism may be what is producing that low-frequency beta hook.

Put out the cat. I'm not kidding—the sudden jolt to the system of an unexpected heavy object landing on you in deep meditation is enough to give the most experienced meditator a beta hook if there is a possibility that it might happen again. Turn the ringer off on the phone (and turn off the audio on the answering machine). Let housemates know not to disturb you—lock your door and put a sign on it if you have to. If you are plagued by a hook into the outside world, these precautions will help you to alleviate it.

But I don't want to give the impression that it will always be necessary to take such elaborate precautions. It would be optimal to be able to sit in the middle of Times Square in New York City and be able to do just as deep a meditation as you could at home. But first we have to learn how.

Sacrificing alpha to get rid of beta: A surefire way of dealing with beta hooks is to go so deeply into trance state that beta is no longer possible. For many people to do this initially, however, it would require almost a loss of consciousness. It is so important for you to begin to experience yourself without the beta that you may be willing for a time to sacrifice your conscious, clear memory of the meditation so that you can get to the depth you are seeking. Lie down to meditate. Don't try to remember the meditation. Use music or (if you have an appropriate one) a guided relaxation to lean on. Go as deep as possible without worrying about staying conscious or remembering content. Once you have learned to really let go of the beta, you can deal with bringing back the alpha waves and once again becoming more consciously aware of your experience.

WHAT IF THERE ARE NOISES THAT DISTURB YOU AND KEEP YOU FROM MEDITATING?

Use them. Anytime that you hear an unwanted sound from the outside, use it to take you deeper inside as you withdraw yourself away from it. I know people who have gotten to the point where they actually welcome a disturbance such as a phone ringing, because it gives them an opportunity to get to a deeper level of meditation. "The sound of the noise is outside—out there. I'm inside myself—in here."

HOW CAN YOU STAY CONSCIOUS WHEN YOU ARE ACTIVATING YOUR SUBCONSCIOUS?

The answer in terms of brainwaves is to keep your alpha active and your beta inactive while you are developing your theta. A way to do this practically (seen, for example, in the House of Doors meditation) is to use imagery to activate the alpha. Hold that imagery present in your mind while you are going deeper and activating theta. For example:

—After clearing your mind and relaxing, visualize a healing room (e.g., meditation room, sanctuary).
—Using as many senses as possible, experience it very clearly. (This will help activate your alpha waves.)
—Find a comfortable place to sit within this room, and allow yourself to go into a deep meditation. (Activate theta.)

Inserting this meditation within a meditation will help you establish the alpha necessary to maintain the bridge to your subconscious mind before entering the deeper part of your meditation.

HOW CAN YOU REMEMBER YOUR MEDITATIONS?

The information above about grounding the experience is very important—i.e., finding keys to remember the experience, bringing them into the beta waves, and verbalizing the content in some way. But you have to have the alpha present as a bridge in order to be able to do this. A period of quiet reflection *about the meditation before you have finished it* will help. In other words, create a definite transition period. You are no longer *in* the contents of the meditation and yet you have not yet completed the meditation. *Go back to the imagery that you originally set up—the healing room, meditation room, or sanctuary. See yourself there once again* (alpha) *and meditate on the meaning of your experience.*

HOW CAN YOU KEEP FROM FALLING ASLEEP?

It is very important not to go into pure delta when you are meditating—in other words, fall asleep. If you are a sleeper, it may help you to sacrifice the depth of meditation for the ability to have a light meditation while staying awake. Then gradually learn to go deeper. Some people fall asleep almost as soon as they close their eyes. If this is you, don't close them! *Meditate with your eyes open, using an unfocused gaze resting lightly on an object.* You may have more trouble producing alpha at first, but this difficulty is worth it if it keeps you from falling asleep.

Sit upright without leaning back against a chair or wall. This alone will probably keep you from falling asleep. The trick now is to learn to go deeply into meditation while sitting upright. Now you need to focus even more strongly on learning to produce theta waves, since you no longer have sleep to rely on as an alternative altered state.

IF I SNORE, DOES IT MEAN I'M ASLEEP?

No, not necessarily. Sometimes in a workshop there will be two or three people who are snoring very loudly. Afterward, someone invariably comments, "O.K., who were the sleepers?" No one answers. No one has experienced themselves as being asleep. If the snorers are pointed out, they may express complete surprise, or they may have been aware that they were making a noise but were too deep in meditation to want to do anything about it.

Snoring is a function of how the air passes through your mouth and nose as you breathe. If relaxation is complete and your body is in a particular position that constricts the passage of air, then snoring is quite likely. It has nothing to do with what kinds of brainwaves you are producing. Most people who find themselves creating this noise while they are producing beta or even sizable alpha waves are uncomfortable about it, so they shift their body position to prevent it from happening. If a person is producing theta without an alpha bridge, he or she probably won't even be aware of snoring because, in a strong, pure theta state, he or she will not be conscious of anything. People producing strong theta and a little alpha may have enough of a bridge to the conscious to be aware of their snoring. Often they are so deep in meditation, however, that they really don't feel like doing anything to stop it because they don't want to alter their state.

Snoring does not detract from the meditation state as long as it does not make you *self*-conscious if there are others in the room. It may bother the other people, however. Use it as a form of biofeedback telling you that you are very deep. If you hear yourself snore, check up on the state of your consciousness. Are you asleep or are you awake? Are you conscious of the content of your meditation? Slightly shift your state if necessary to be more where you want to be.

HOW CAN YOU KICK START YOUR ALPHA WAVES?

Producing alpha waves is physiologically linked to your eyes. As we have seen, if you close your eyes, you will produce alpha quicker and stronger than if you keep them open. *If you roll your eyes upward in your head, you will produce alpha even more.*

Try this exercise:

—With your eyes closed, inhale and gently raise your eyes to look toward the center of your forehead.
—When you exhale, lower your eyes.
—Repeat several times.
(Note: Do not strain your eyes. If you feel any discomfort, stop immediately.)

Now try it this way:

—With your eyes closed, exhale and gently raise your eyes to look toward the center of your forehead.
—When you inhale, lower your eyes again.

Most people find the first method more natural and therefore more effective. However, a few have difficulty with the first method but swear by the second. Try both and you will know which works for you.

I suggest you use this exercise sparingly. After a few minutes, its alpha-producing quality diminishes. Also, the eye muscles are very delicate and easy to tire, and creating eyestrain would be counterproductive to what you are trying to do. A few breaths—five to ten at the most—at the beginning of the meditation would be most effective.

A NEW GATEWAY TO AN ANCIENT PATH

The state of meditation is nothing new—only these descriptions of it are. With the development of modern technology, we now have available the means of measuring and monitoring the meditation state as never before. In our Western world, people sometimes want to use technology such as an EEG to determine the validity of the masters, to discover whether they are really in the state they claim they are in. My travel in the Far East has shown me an interesting reversal of this mind-set.

I was teaching my work to a *chi kung* class in a Taoist temple in Taiwan. The sea of faces in front of me was disbelieving and resistant, even hostile. Getting the information across was tricky because of the language barrier and the need for constant translation. Also, I was bringing a machine into their sacred sanctuary! On top of that, I was a woman!

I introduced the equipment and the theories, drawing pictures of the brainwaves, including the awakened mind pattern, on a blackboard. I asked for a volunteer to be connected to the Mind Mirror and was met with silence. The *chi kung* master looked around the room and saw that no one was going to come forward, so he raised his hand and volunteered. His students looked at him in concern and awe.

I was more than a little nervous at this point, because I wanted this lecture to go right, yet there was nothing I could do to ensure its success. I certainly couldn't guarantee that the master had the right brainwaves.

I hooked him up and, with great relief, was able to point to his perfect awakened mind pattern showing on the screen. Then I pointed to the picture of the awakened mind I had drawn on the board to show that they were the same.

The group of reserved and resistant students broke into smiles and laughter. As if one body, they moved forward to surround the machine, to look at it, touch it and see it for what it was. They all wanted to be the next one to try it.

They had used the master to validate the technology rather than the other way around. Only when they saw his perfect awakened mind pattern did they know the Mind Mirror worked.

IV

MEDITATION
AND THE MATERIAL
OF THE MIND

There are two ways of looking at consciousness—*state* of consciousness and *content* of consciousness. So far, we have primarily addressed the state of consciousness as determined by brainwave frequencies. But what about the content? What is the actual material of the mind? How can we work with it, understand it, direct it, and use it?

The material of the mind takes several different forms, depending on which brainwave state it is being experienced in. For example, our *thinking processes* occur within the container of our beta waves. When we remove the material of the mind by quieting our thoughts, we create a clear and empty field for the *content* of the alpha and theta waves to arise in. This content may include sensory images, daydreams, repressed memories and emotions, simple awareness of whatever is, or inspirational insights. When we add useful and appropriate beta brainwaves back into the alpha-theta meditation pattern, this creates an awakened mind brainwave pattern, fertile ground for such content as creativity, problem solving, and self-healing. The awakened mind brainwave pattern with this kind of valuable, beneficial, and constructive content gives us the high-performance mind.

The material of the mind can be both helpful and bothersome. In the next few pages we will look at how to deal with some of the more bothersome types of content—the static of beta that hinders deeper states of consciousness. Then we will explore the mental content that occurs at the border between the conscious and subconscious in the alpha brainwaves, and finally, the substance of the sub-conscious that occurs in the theta waves.

REDUCING THE CHATTER OF BETA WAVES

Many of our ordinary waking states contain too much high-amplitude beta—or what we earlier called splayed beta. One of the most vital aspects of learning brainwave mastery is to get these often unruly thought forms under control.

The following exercises have two benefits—they help reduce beta brainwaves and they are *short*. Even after only one minute of meditation, you will feel less stressed and more centered. Practice these at any time, but especially practice these when your mind seems to be filled with thoughts that you would rather not have.

ONE-MINUTE MEDITATIONS

Stop and sit for one minute. Do not try to alter your thoughts or control your mind in any way. You can, if you wish, time yourself on a watch or clock.

What happened to your mind during that one minute?

—How many thoughts did you have?
—Where did they come from?
—What was the *content*, or subject matter, of your mind during that minute?
—How many times did you change the subject?
—Were you completely aware of your thoughts as they were happening, or was there a running commentary somewhere in the back of your mind that you could tune into or tune out of?

Can you imagine what the experience would have been like to spend that minute with a *still* mind?

RELAX YOUR TONGUE

Sit again for one minute. This is best done with your eyes closed. You can either use a timer or simply estimate when you feel one minute has passed. You'll probably be quite surprised with your estimation the first time you try it.

Allow your tongue to relax.
Focus *only* on relaxing your tongue—nothing else.
As you relax your tongue, you may feel it shorten and thicken.
Imagine that it is floating in the cavity of your mouth.
Exaggerate the relaxation of your tongue.
Allow your chin to drop, your jaw to loosen, and your mouth to open slightly.
Each time you exhale, allow your tongue and the whole mouth area to relax even more.
Do this for one minute.

If you have not used a timer, look again at your clock or watch. How much time has actually passed? What happened to your mind during this minute?

—Did a minute seem slower or faster than the first time?
—How many thoughts did you have?
—What was the content of your mind this time?
—How many times did you change the subject?
—What do you think happened to your beta brainwaves?

Simply by relaxing your tongue, you can reduce the activity of your thinking mind and begin to let go of unwanted thoughts.

Try relaxing your tongue again, but this time, after ten or fifteen seconds, *think.*

—What happens to your tongue?
—Does it become more tense?
—What happens to your thought process?
—Is it more difficult to think while keeping your tongue relaxed?

Experimenting with this meditation in different ways will show you that relaxing your tongue can be an invaluable key to stilling your mind. It is a simple technique that will help you radically reduce your beta brainwaves and begin to still your mind immediately. When people think, they have a tendency to talk to themselves. Even if you are not aware of this subvocalization, your tongue will feel tense—ready for action—when you are thinking. You may experience this as a gentle pulling on the back of your tongue or you may not feel it at all. If you relax your tongue *completely*, you cannot talk to yourself. Therefore, it is much more difficult to think.

Let's look now at other ways of stilling the mind, turning off the beta so that the mind can more readily experience the benefits of the remaining brainwaves. One means of accomplishing this is to give the beta waves something upon which to focus. In this way you can teach yourself to develop a *one-pointedness of mind* through concentration.

HOW OFTEN DO YOU BREATHE?

One of the most effective and easily available focal points on which to concentrate is your own breathing. Try another experiment.

Time yourself for one minute, and count how many breaths you have during that time. Do not try to alter your normal breathing rate. One complete breath cycle of inhalation and exhalation is one count.

The average number of breaths per minute is between twelve and eighteen. What is *your* normal number of breaths per minute? If you concentrate on your breathing and allow it to slow down, you will instantly reduce your beta brainwave activity.

SLOW DOWN YOUR BREATHING

Lightly place your hand palm down on your thigh. As you inhale, slowly raise your hand. As you exhale, slowly lower it again. In this way, simply use your hand as a lever to gently begin to regulate your breathing.

It is important to make sure you are not forcing yourself to slow your breathing. Nor should you hold your breath or in any way starve yourself of oxygen. You simply have an *intention* of slowing your breath. You may wish also to linger a little longer than usual at the points of full inhalation and full exhalation. However, there should be absolutely no strain.

Once again, time yourself, and allow yourself to count the number of breaths that you have in one minute, this time intentionally slowing your breathing rate in the manner described above.

Afterward, notice the differences between the first minute of counting breaths and the second.

—How many breaths did you take this time?
—Was it fewer than before?
—How long did the minute seem?
—What were your thoughts like, compared to the minute you timed earlier?
—Did you remember to keep your tongue relaxed?
—How do you feel?

Notice what only one minute of slowing your breath can do to your sense of stress or relaxation.

CONCENTRATION EXERCISES

Learning to focus your attention on one subject without letting your mind wander will greatly reduce beta waves. It will also expand your awareness and make achieving a meditative state easier.

Begin this exercise sitting with your eyes open. Choose a spot to look at a few feet in front of you at slightly lower than eye level. You may place something in front of you or use an already existing ordinary object in your room. People often choose a flower, a design such as a mandala, or an icon of some sort for the aesthetic value, but the object of the exercise is to gaze at an unidentifiable spot, so if you choose something like a flower, identify one spot on the flower to look at—not the whole flower. I know some people who are completely satisfied gazing at a spot on the carpet.

You can use a one-minute timer for this exercise if you wish, or simply practice it for about a minute. Sometimes you may be surprised as your minute stretches into a longer time period.

Sitting comfortably with your back straight, send your awareness outward to the spot you have chosen.

Become aware only of the spot. Let your awareness move outward in a straight line, from you to the spot.

Let your eyes relax and unfocus your vision. The spot or the area around the spot may become hazy or clouded. (Remember, you are not focusing on an identifiable object—only on a location.)

Allow your mind to clear of all thoughts and images as you focus only on your spot.

Afterward, reflect on your experience.

—How many thoughts did you have?
—How long did the minute seem?
—Were you able to maintain your focus?

Some people prefer to focus on something within instead of something outside. If you feel more comfortable with this technique, choose a spot *inside* on which to concentrate. Favorite areas among the people I have worked with include the heart area, the navel, the bridge of the nose, the center of the forehead, or the crown of the head. Your spot, however, could be anywhere inside that is comfortable—even your left big toe. Practice this exercise with your eyes closed. A final form of these exercises involves learning how to focus inside and outside at the same time.

RENATA'S EXPERIENCE

Following is the account of one student's response to these concentration exercises. Her experience took place in a group setting in Big Sur, California.

At first I sent my focus to the rock at the fireplace. As my focus became stronger and stronger, everything in my peripheral vision seemed to look like a negative. Suddenly I began seeing the aura of red around the person closest to this rock.

The next exercise of focusing on the inside was more difficult for me, but not impossible. I realized how I had neglected my stomach (since that was the point I chose to focus on). It seemed to have been nonexistent until this moment.

The final exercise of focusing on the inside while simultaneously focusing on the outside was the most incredible experience. Suddenly there was a merging between the rock and myself, and all I could feel was unity and love—no more boundaries, no separation. Fascinated by this experience, I continued this practice throughout the day while interacting with people.

Renata's experience teaches us that meditations do not have to be elaborate or done over a lengthy period of time to have a positive effect upon our perceptions. Renata underwent a fundamental change in her perceptions of her relationship to other people after practicing these simple focusing exercises. She goes on to say:

> My perceptions have changed completely by having discovered this tool. By focusing on the inside while focusing on the outside or the person I was interacting with, I once again felt only love. All superficial judgments disappeared.

Many responses to these exercises are possible. Try not to have particular expectations, but simply allow yourself to experience the possibilities.

MASTERING BETA WITH IMAGERY

Images can also be used to master the beta contents of your mind. There are many types of thoughts that may occupy your mind unnecessarily, ranging from the irrelevant to the deeply stressful. There are several imagery techniques that will help you master unwanted beta. I have listed some below.

1

Visualize the thoughts as gray clouds in a blue sky. See this clearly. Any thought that comes into your mind is a puff of gray cloud in an otherwise clear sky. Gently begin to blow away the clouds with your breath as you exhale. Continue blowing the clouds away until the sky is completely blue and clear.

2

Occasionally there are thoughts in your mind with emotional content that you would like to eradicate completely. Imagine writing these thoughts on a piece of paper, then throwing the paper into a fire and burning it. Notice what you feel as the thought burns up.

3

Sometimes you may have a creative thought or an important idea that keeps returning when you are trying to still your beta brainwaves. Because you don't want to lose the contents of this thought, you allow it to continue to distract you. If there is a thought in your mind that you would like to store in some way, imagine writing this thought on a piece of paper, then filing it away in a briefcase or filing cabinet where you can store it to refer to later whenever you wish.

You can notice from these examples that there are many kinds of imagery that can be used to help you clear the contents of your mind. You may wish to try making up your own images to help you master your beta brainwaves.

WITHDRAW FROM YOUR THOUGHTS

Withdrawal into yourself in meditation is another excellent means of leaving unwanted thoughts behind. Get into a comfortable position, allow your eyes to close, and take a few moments to relax again.

Send your awareness outside of the building you are in—into the street. Become aware of external noises. You may also be aware of the energy and atmosphere outside the building, especially if there is traffic or people moving around. Try not to label the noises or energetic vibrations, but just be aware of any sounds that you may hear.

Gently withdraw yourself from the outside environment into the building. Try to take in the whole building, no matter how large it is. Become aware of any sounds, energies or movement within the building.

Now gently withdraw yourself from the rest of the building into the room that you are in. Become aware of the sounds, energies, and atmosphere of the room that you are in.

Now gently withdraw yourself from the room into your own body—into yourself. Become aware of the sounds, sensations, movements, and energies within your own body.

Now finally withdraw yourself from your body into a place inside yourself where there is stillness—into your center. There exists inside you a place of stillness, silence, and peace. Find the place inside that is quiet and allow yourself to rest in that place for several minutes.

Anytime a thought comes through, use one of the techniques that you have learned in this chapter to release it, and return to that place of stillness.

As you prepare to return to the outside space, very gradually begin to be aware of what you are experiencing—the feeling of relaxation in your body, the contents of your mind—and once again be aware of the place of peace and stillness inside.

You may wish to find a few keys for what this feels like to help you return to this state of consciousness more easily and effortlessly. These keys or cues might be a body sensation, a color, a word, an image, or even a sound.

After doing these exercises, take several deep breaths and allow yourself to stretch and reawaken fully, feeling alert and refreshed. Make sure that you are completely back in the outside environment before you begin dealing with it. In other words, these exercises can take you deeper into yourself than you are consciously aware of. Don't drive a car or operate any dangerous equipment without making sure you are fully back.

THE MATERIAL OF ALPHA
Accessing Alpha

The next step after you have learned to control your beta is to learn how to access alpha brainwaves. Many people produce some amount of alpha as soon as they close their eyes. Alpha is certainly much easier to produce with the eyes closed rather than open. One exception to this rule can be visual artists. I have found that those who use their eyes to accomplish their art in a meditative state are sometimes more able to produce open-eyed alpha when they wish. This could include painters, sculptors, writers, choreographers, photographers, and so forth. However, their alpha is still usually not as strong as it is with closed eyes.

Our goal at present is simply to practice producing as much alpha as possible. So close your eyes for the following meditations.

Visualization and various forms of imagery help access alpha brainwaves. It is important to realize that imagery takes place in all of the senses—not just the visual. Using as many senses as possible when you are visualizing is vital to the production of alpha brainwaves. Relaxed, detached awareness and deep relaxation also help us access alpha brainwaves.

By itself, alpha is a form of mental relaxation and light reverie, which is stimulated by all forms of sensory imagery. Alpha is that state between waking and sleeping where vivid and sometimes unintelligible images pop into the mind unbidden. It is also the place of daydreaming, where the mind wanders off into images of the past, the future, or fantasy. It is the gateway to meditation, where the scene is set, the groundwork is laid, and the bridge is opened to contact the deeper reaches of the mind. Adding alpha to your brainwave profile will promote better memory, greater retention of creative insights, and more useful awareness of deep, subconscious resources.

There are two main ways of experiencing the content of alpha brainwaves. The first is almost an absence of content experienced as a relaxed, detached awareness of whatever simply *is*. If a clock is ticking loudly in the room and a window is open, providing a cool breeze, the content of the mind is simply that—the awareness of the clock ticking and a cool breeze.

Often, however, there is some form of imagery involved in the material of alpha. Alpha waves give us our clearest, sharpest, and strongest visualizations.

VISUALIZATION

The concept of visualization has become very popular in recent years. The adage "Visualize what you want to have happen, and it will happen" has been drummed into us from all sides. While there is much truth in this, a far more complex process must take place before the benefits can be truly reaped.

"BUT I CAN'T VISUALIZE!"

Many people have fixed ideas about what visualization means, how to visualize, and whether or not they are doing it correctly. Some are so discouraged because they think that they can't do it that they stop trying. They begin to believe that visualization just doesn't work for them. Others do it, but secretly feel they are doing it oddly, or try to limit their experiences to be within the acceptable range. Still others experience the imagery but discredit its value because "it's only visualization and doesn't mean anything." Often the fear of being incapable of visualizing is based on an inadequate definition of what visualization actually is. The next section will place visualization in a broader context, showing how this technique is accessible to everyone. Read the next paragraph through and then follow the instructions.

Close your eyes. Imagine that you are going to have a drink of some sort. Is it in a cup or a glass? What color is the container—or is it transparent? What color is the liquid? Feel the container. Is it hot or cold? How full is the container? Pick it up and feel the weight. Put it back down on the table heavily. Can you hear the bang? Smell it. Now, finally, taste it. Be careful not to burn yourself if it is hot. Drink as much as you want. Open your eyes and return.

What happened? What were you drinking? Could you taste it or smell it? Could you see it? Could you feel or hear the container? If you can answer yes to any of these, you can visualize, even if you had trouble seeing it with the visual sense.

The connotations we have come to give the word "visualization" have done a grave injustice to our mental imaging processes. Most people believe that in order to visualize, they need to have the ability to have visual imagery, almost as if they were watching scenes on a mental movie screen. They believe they cannot visualize because they do not have the capacity to see images in quite this way.

What they do not realize is that they have five other senses with which to visualize—hearing, touch, movement, smell, and taste. It would serve us much better here to change the word "visualization" to "sensualization." Each one of us has an *access sense,* the sense that we automatically use first. Since it often provides us the strongest image content, we usually assume that this is our *only* content. Not so. Starting with our access sense or any sense we choose, we can build a much fuller picture of any scenario we are trying to sensualize by adding input from each of the other senses. All of our old suppositions about visualization can become enriched and enlarged by using sensualization.

It is beneficial to *practice* the art of sensualization and to increase the clarity and vividness of the material of the alpha waves. You will find that practicing sensualization not only helps you improve your sensualization ability; it also increases and improves alpha brainwave production. Practicing the kinds of imagery presented in the following exercise will help you develop greater mental flexibility. The stronger the image you can evoke, using as many of your senses as possible, the more successful will be your outcome, regardless of your purpose. For the best results, practice these exercises while sitting or lying in a relaxed position. You may wish to read them into a tape recorder so that you can listen to the instructions and keep your eyes closed throughout. Otherwise, read one instruction, then close your eyes and imagine it as clearly as possible. When your mind starts to wander or you feel your imagery is as complete and as real as it can be, open your eyes and read the next instruction.

SENSUALIZATION EXERCISES

SEE

—Begin by seeing as many colors as possible, not only the primary colors of red, yellow, and blue, but also the secondary colors of orange, green, and purple and their various combinations, shades, tones, and tints—light blue, aqua, brown, pink, crimson, gold, peach, lilac . . . Then add the opposites of black and white.

—Watch the clouds moving across a windy sky.

—See the vastness of thousands of stars in the sky on a summer's evening.

—Light a match in a darkened room.

—Standing on top of a high mountain, gaze over miles and miles of beautiful countryside below you.

—See a friend's face smiling at you.

—See the street your home is on, quickly passing each house until you get to your own front door.

—From your window seat in a landing airplane, see the city, the buildings, the streets, the cars, the airport, and finally the runway, as you get closer and closer.

TOUCH

—Feel soft fur under your fingers.

—Walking in the park, run your hand over the rough bark of a tree.

—Caress the soft smoothness of a baby's skin.

—Take ice cubes from the freezer and feel them melt in your hand.

—Wash your hands with soap and warm water.

—Experience the gritty feeling of brushing sand off your legs at the beach.

—Imagine the way your scalp feels as you comb or brush your hair.

—Experience the cool air on your steaming body as you step out of the hot bath.
—Feel a roughly cut plank of wood, but be careful of the splinters.

HEAR

—the sound of a jet taking off
—the *whap* of a tennis ball against the racket
—children laughing at the playground
—the dentist's high-speed drill
—the cheerful crackle of wood burning on an open fire
—a foghorn echoing in the distance
—the howl of a police siren
—the dawn chorus of birds waking you in the early morning
—laughing voices, clattering cups, and tinkling glasses at a party

SMELL

—the potent fumes you experience driving behind a truck
—the smell of bacon frying
—the soaps-and-lotions department in a store
—the smell of a seafood market
—the smell of burnt toast
—your favorite perfume or aftershave
—the smell of bread baking in a bakery
—the smell of freshly brewed coffee
—the smell of shoe polish as you are polishing your shoes

TASTE

—the sharpness of pure lemon juice
—the milky foam on a cup of cappuccino
—a menthol cough drop
—the salt on a potato chip
—a bite of sharp cheese
—a crisp, fresh apple
—cold vanilla ice cream melting in your mouth

—a large, ripe strawberry
—the minty freshness of toothpaste

EXPERIENCE (KINESTHESIA)

—the sensation of walking downstairs in the dark and feeling for the light
—reaching up on a high shelf for a container you can't see
—walking barefoot on a pebble beach
—playing tennis, baseball, volleyball, or your favorite sport
—dancing
—struggling to open a childproof medicine cap
—sunbathing, or just lazing on a sheltered, deserted beach
—receiving a warm hug from an old friend

Which sense provided you with the most vivid experience? Which was the weakest? Following is a scale on which you can rate your experiences in each of the categories of senses.

SENSUALIZATION SCALE

1—Lifelike imagery, as vivid and distinctive as the actual experience
2—Quite clear and well-defined imagery, representative of but not quite equal to the actual experience
3—Intermediate clarity, but still realistic imagery
4—Hazy imagery with minimal representation of the actual experience
5—Formless, dim, indistinct and unrecognizable
6—No imagery at all, only thinking of the experience

Did the strongest images all occur in one or two categories, or were they evenly spread throughout the senses? Did the strongest images come in brief flashes that disappeared as quickly as they came? Or were you able to maintain the vividness for a consistent period of time?

Do not discount even the briefest imagery flashes. Often people will not recognize or admit to themselves that brief instant of insight when the experience was extremely lucid. They continue to think that they can't achieve clear mental imagery simply because *they are not able to maintain it*. If you can recognize even that split second of clarity, you can learn to lengthen it much more easily.

Was there one sense that always occurred first, before the others were able to appear? This is your *access sense*. Often we use it without realizing it, to activate the other senses in the imagery process. For example, you might see the jet first, before you hear it taking off. Or you might feel yourself strike the match before you see the light it emits, or smell the cough drop before you taste it.

Sometimes people are stuck in only one sense—frequently it is visual. Notice, when you tried to *hear* the bell ringing, did you *see* it instead? When you wanted to *smell* the bread, did you find yourself *looking* at a loaf? When you have one sense that is overpowering the others, you can use it as the foundation of your sensualization; you can build on it rather than stopping with it.

Try the exercise again, but this time use your strongest sense to help you access your weaker senses. In other words, if your strongest sense is visual and your weakest is auditory, see the bell before you hear it ring—but don't just stop at the seeing. If your strongest sense is kinesthetic and your weakest is visual, feel yourself go through the motions of striking the match before you see the light flare up. If your strongest sense is auditory and your weakest is tactile, hear the sound of the ocean at the beach before you brush the sand off your legs. Play with these combinations of senses and images in any way that will help you improve the quality of your mental material. Remember that, ultimately, the more senses you can combine to create a vivid internal experience, the more productive and meaningful your sensualization will be.

As you are developing your imagery ability, you are also improving your alpha brainwave production. This means that you are improving the bridge to your subconscious mind. The more you practice, the better your abilities for recalling theta contents will become. This will improve dream recall, meditation recall, and memory access.

EVERYTHING YOU ALWAYS WANTED TO KNOW ABOUT VISUALIZATION BUT WERE AFRAID TO ASK

The following are some of the most common questions and misconceptions that people have about visualization.

"WHERE DO YOU EXPERIENCE YOUR VISUAL IMAGERY?"

The majority of people experience their visual imagery somewhere in front of their eyes, for example, as if it were projected on a movie screen in front of them or being acted out live around them. But many other people do not see visual imagery in this way. Other locations where imagery is seen are in the middle of the head behind the eyes, in the back of the head, outside the body in back or off to the sides, or inside the body (not in the head) in the heart, solar plexus, or stomach. It is also possible to have multiple visual fields in two or more distinct locations with two or more different imagery scenarios taking place at the same time. I have a client who has one visual field for yes and one visual field for no. The two fields change size in accordance with the strength of each attitude.

"CAN ALPHA BE USED FOR BETA FUNCTIONS? FOR EXAMPLE, CAN IMAGERY BE USED TO DO MATH?"

Demonstrating the Mind Mirror in a class one evening, I asked the man who was hooked up to do a math problem in his head, fully expecting to show my students the normal beta that is usually stimulated by this type of problem solving. Instead, this individual produced copious amounts of alpha waves and very little beta. Somewhat surprised, I asked him how he did his math. With a slight flush of embarrassment he said, "Well, you know, every number has a color, and when you add orange and green you get the answer."

Another individual who utilized alpha to solve math problems described his experience in terms of perspective. His system required that certain numbers be on the right, certain numbers on the left, other numbers at a distance, and specific numbers up close. He used a very complex visual formula to work out his math problem. You may be more familiar with the quite common practice of visualizing the numbers on the blackboard or writing them on a piece of paper in your mind.

Perhaps the most interesting and unique response I had was from a young woman who produced very strong *theta* while figuring out the answer. When asked how she did her math, she explained that, as a child learning math in school, she had needed glasses but no one knew. Because she could not see the blackboard to read what her teacher was writing, she said she had learned her math by reading her teacher's mind!

"WHEN YOU SENSUALIZE, ARE YOU OUTSIDE THE IMAGE LOOKING IN, OR INSIDE THE IMAGE LOOKING OUT?"

You can be either. Some people experience the image as if they are in their own body using their senses in a very real way. When they see visual imagery, it is as if they are looking out of their own eyes at the images. When they feel a tactile image, it is as if they reach out their hands and touch it. Other people, mostly strongly visual people, see the image as if they are looking at themselves doing the activity. If the guided fantasy, for example, instructs you to take a walk along a country path, they see themselves walking along the path.

So is one way better than the other? Not necessarily. There are times when each has a function. If you want to use your imagery to make a change within yourself, the more you can experience it being inside you, the more effective your imagery will be. On the other hand, sometimes we use imagery to explore something that is dangerous or frightening, and if we were to experience it directly through the image inside ourselves, we would not be able to stay with it. By distancing ourselves from the image, we are better able to control our reaction to the experience and therefore work through the whole process.

I have even had clients who put a clear Plexiglas wall between themselves and the image, further distancing themselves from a scene that is too painful or dangerous for them to experience directly. This serves to give them a sense of safety not present if they are inside or right next to the image. They can then finish the process, completing their imagery experience rather than abandoning it in fear or pain.

The optimum situation would be to have the ability to be both inside and outside the image at the same time. This may sound paradoxical to you, but not only is it possible, it is actually common for some people. In order to develop your mental flexibility, try practicing each imagery style separately, and then try them together.

Let's go back to your walk along that country path.

Visualize the scene and see the vegetation on either side of the path. See the colors and shapes. Feel the textures with your fingers. Notice the time of day, the weather, the temperature, the smells and sounds. Experience this as if you are inside your own body looking out.

Now stop and start the exercise over. This time *see yourself* walking down the country path. Place yourself in the picture and look at it from the outside.

Was one of these methods easier than the other? Practice both so that they become familiar. Now see if you can combine the two. At first you may find yourself switching back and forth. Then the switching back and forth becomes more and more rapid. Finally you experience both perspectives at the same time. Having the mental flexibility to move around within your imagery in this way will give you the maximum results when you wish to use your imagery to accomplish something.

''WHAT ARE THE DIFFERENCES BETWEEN ALPHA IMAGES AND THETA IMAGES?''

Images generated by alpha waves are clearer, sharper, and usually more vivid than those generated by theta waves. The colors are stronger and more varied. The edges are cleaner. Images generated by pure alpha are also less meaningful and do not seem to be coming from deep within you. Like the exercises described above, they are easier to create on command.

In contrast, images generated by theta waves are usually darker, fuzzier, and less distinct. They can be intensely personally meaningful and feel as if they are coming from the very depths of your being. Or they can feel as if they are divine inspiration, a gift from some sort of higher power. Sometimes they seem to be surrounded in light. A sense of knowing often accompanies them—knowing that this is the truth, knowing that this is the right action to take, knowing that you have touched something important inside. But don't always expect clear insight and moments of illumination from your theta waves. You can also swim around in your subconscious as if it were a murky swamp of perplexing obscurity.

Occasionally theta imagery will be preceded by or accompanied by washes of blue or purple colors. These signify a strong flare of theta waves. Have you ever heard the saying, "It came to me out of the blue?"

"HOW CAN I IMPROVE MY IMAGING ABILITY?"

Practice, practice, practice! Utilize the exercises above to practice on the senses that are your weakest. Find times during the day when you can close your eyes and sensualize. In your morning shower, when the water is running over your face and your eyes are closed, take the opportunity to sensualize the day in front of you. Imagine with all of your senses what you would like to have happen. If you have a problem or a question that needs to be answered, let the solution come to you through imagery. Leave the words out. Don't talk to yourself. Let your images do the communicating.

Don't tell yourself that you can't visualize. This negative reinforcement is very detrimental to the imagery process. Validate each image as it comes. When you have those brief moments of clarity and lucidity, recognize them and appreciate them, no matter how short they are. If you do that, they will inevitably return for longer periods.

"IS THIS IMAGE REALLY COMING FROM ME, OR AM I JUST MAKING IT UP?"

This is a frequent question that causes great concern to many people. Actually, the answer is quite paradoxical. *Anything* you make up comes from you; some things simply originate in different parts of your mind than others. Some images arise spontaneously out of the subconscious, other images are generated by the conscious mind. Compare the qualities of your images to my descriptions above to help determine where each image is coming from.

Trust your imagery! In working with the content of consciousness with clients in my private practice, I find that often people may have a theta-generated image of some painful part of themselves communicating a message as to why they are undergoing certain difficulties. For example, a client remembers her pain and confusion as a little girl when her mother abused her. Because she had always been told that she had a perfect childhood, she has a defense mechanism set up to deny the validity of these subconscious images. She sees the abuse in her mind and asks me, "Is this real, or am I making it up?" If she is seeing it and feeling it in images coming from her theta waves, then it is valid. Even if the child condensed some images and eliminated others, the essence of what she is reexperiencing is real *within herself*. This does not necessarily mean that her mother abused her in the way she remembered. But it does mean that she has experienced abuse inside and needs healing for that.

Even if she were to intentionally make something up, the material that she used to make up the experience had to come from somewhere inside her mind. Therefore, what she is sensualizing is worth approaching as *her* experience.

THE MATERIAL OF THETA

When people alter and develop their states of consciousness through meditation, intense emotions and memories may arise. People have a diversity of positive and negative reactions when confronted with the unexpected contents of their subconscious minds. What they do with that content often depends upon the needs, experience, and psychological makeup of each individual.

If you are confronting hidden traumas and feel you are getting in over your head, I urge you to seek the help of a good mental health professional. Getting proper support while you are clearing out painful subconscious material may be necessary to assure that the process is effective, lasting, and safe.

Educating yourself about theta waves and the subconscious and taking steps to *actively access* this previously hidden or elusive content is the most direct and effective way of dealing with it. There are a number of guidelines both for accessing this material and for dealing with it in a healthy, successful way.

But first, let's look at some examples of people accessing subconscious content. We will use the House of Doors meditation from Chapter Three as an example. When you open the door and see what is behind it, you are stimulating theta waves and releasing material from them. There are endless possibilities as to what you might experience, just as there are an endless number of theta doors to open. Here are some examples from my own sessions with clients.

Content: Behind the door is a room that has not been opened or seen the light of day for a long time. Dingy and dark, it is filled with thick dust and cobwebs and old deteriorating furniture. It is hard to breathe there.

What she did with it: To the meditator this represented her subconscious as a whole—a place she had not visited for a very long time. She threw open the windows and added new ones that weren't there before, thoroughly cleaned the room and rearranged the furniture, and turned the room into somewhere comfortable to be. She now has a place in her mind to go and meditate when she wants to quickly access her subconscious.

Content: Behind an antique brown door, a man finds his dying father in a hospital room.

What he did with it: Even though his father had died years before, this man realized he had never made peace with him. Therefore he had unknowingly lodged this unfinished business and the pain of his father's death in his subconscious. In his meditation, he cried with his father on his deathbed. He asked his father's forgiveness for wrongs he had done and told his father he equally forgave him. He held him while he died, something he had wanted to do but had not been able.

Content: A woman opens a green door to find, not a room, but a wide open expanse of beautiful countryside.

What she did with it: She realized from this meditation that nature is a part of herself that she had been denying. She lived in a city and spent her days in an office and had very little contact with the outdoors. She realized that she needed to give herself more time in nature.

Content: Behind the door is pitch black darkness. Try as she might, the client can see nothing. She leaves with discouragement.

What she did with it: I asked her to go back and find a light switch. It was as simple as that. She turned on the light and found a room from her childhood where there were very painful memories. She was then able to begin to deal with those memories in a healthy and conscious way.

Content: The label on the door says LOCKED. The meditator tries as hard as he can but he cannot open the door.

What to do about it: The subconscious really wants to protect the content that is in there. My suggestion in these cases is to leave the door alone for now. Instead, look down the hallway for a door marked KEY. Go inside and see what you can find out about why the other door is locked.

Content: The label on the left-hand door says DEATH and on the right-hand door says LIFE. The images behind each door fit their labels.

What it means: The doors sometimes come in pairs of opposites like this—dark and light, happy and sad. Simply explore the rooms to find out more information about your own subconscious.

Content: Behind a red door a fire is raging—a wall of flame.

What she did with it: As the meditator stepped into the fire, she realized this was the anger she had kept hidden away for years. From this she realized she needed to start dealing with her anger and learning how to express it appropriately. This room of fire became a place she enjoyed returning to over and over again to contact her deeper feelings.

Content: The room is an exquisitely appointed meditation room. It is serene, calm, and peaceful, and needs no changes.

What she did with it: She found a place of serenity inside herself where she could return to meditate and find peace.

A SKEPTIC'S STORY

One of my students, a man I will refer to as J.T., felt very skeptical about the whole process of meditation. Initially resistant to participation in the workshop, he expressed doubt that much could happen inside himself while he was hooked up to an EEG in a roomful of strangers. When he tried the House of Doors meditation, he was pleasantly surprised to find out that he had been wrong. His meditation illustrates how individual and idiosyncratic both the content and the interpretation of its meaning can be.

His house was a chapel and the multiple-mirror section revealed an image of himself dressed in a long white robe. The first door he entered led to a six-cornered side chapel. The second door led to what appeared to be a medieval torture chamber with a stretching table, which transformed into a deathbed. This bed depicted to him his potential laziness and lack of drive on his spiritual path. The message that he received was that he no longer needed to be afraid of hardship and death. With great emotion, he explained how touched he was by the love in this message. After this meditation, he was truly moved by the depth of his experience and the power of the messages from his subconscious. He was also impressed by how unintrusive the EEG equipment had been.

DEALING WITH THE CONTENT

The following are principles you can use in dealing with the material of the theta waves:

Don't be afraid of the content. Whatever appears is part of you, and as such deserves to be respected and even loved, no matter what it is.

You can always make changes inside. Nothing is permanent or fixed about these images and the inner states they represent. You can take this information and act upon it.

You don't have to do the work all at once. Understanding, healing, altering, and integrating your issues may take time.

You can always return. If you found these images once, you can find them again. Try to leave markers, or find signposts to help you reenter the state of consciousness where you found the content you want to revisit.

Nothing is too overwhelming or too painful to eventually be healed. Perseverance and willingness to heal have a remarkable effect on the subconscious.

If you don't find anything significant, don't worry. Some people are afraid that they are not really in theta unless they are experiencing something deep, meaningful, creative, or difficult. Absolutely not! Theta can be a place of quiet contemplation and rest.

Develop your spirituality. Creating or strengthening a conscious connection with whatever divine source or higher power you believe in will give you limitless resources for healing, growing, and grounding in your theta.

INTENTIONALLY ACCESSING SPECIFIC
SUBCONSCIOUS MATERIAL

Chapter Three gave some specific guidelines for accessing theta brainwaves, but the House of Doors meditation is designed only to access generic theta. Anything could happen inside the rooms because we aren't looking for content about a specific issue. What if we want to access specific subconscious material?

There are many different ways into the subconscious. Again, using imagery is perhaps one of the most direct and effective tools, simply because the imagery opens the alpha bridge as you go down and allows you a better and more immediate conscious awareness of the content. If we want to access specific material, however, we need to find imagery that can guide us to what we are looking for.

A meditation like the House of Doors is easy to convert from the generic to the specific. First, you must decide the category of material in which you are interested. Try to find one image or word that describes it. For example, instead of the question "Why I am having a difficult time stopping smoking?" you might want to have an image of a lit cigarette or the word "smoking" in your mind. If you are interested in exploring your relationship, instead of having "What can I do to improve my relationship with Jane, and how can I make it more the way I want it to be?" in your mind, use "Jane" or "relationship." Depending upon the problem you are experiencing, you might prefer something like "communication" or "fulfillment," or simply the image of a heart.

Once you have found a word or symbol for the material you want to explore, go back to the House of Doors meditation. When you are walking down the corridor passing doors, find the door with that label on it. Open *that* door and explore what is behind it.

Depending on what you find inside, you may want to bring up the more detailed issues once you are in the room. But you need to take it one step at a time and let the subconscious material take the lead. If you open the door with the cigarette on it and find an image of yourself inside, you may want then to ask that image what he or she needs to do to stop smoking. If, on the other hand, you open the cigarette door and find a hospital room, or the grim reaper, you may have a different series of images. Let the information from the sub-conscious take the lead. Do not try to control *how* the information comes to you. You can still follow the principles above for making changes inside—but it helps to know what is going on inside that you want to change.

It is not uncommon for people to have a lack of theta brainwaves because some event or events in their past are painfully lodged deep within the subconscious and the psyche has decided to lock them away from conscious knowledge. It can be fascinating to watch the brainwaves of someone who has subconscious material that wants to remain hidden.

It is possible, while looking at brainwaves on the EEG monitor, to actually watch the content being shifted around to prevent the con-scious mind from becoming aware of it. Figure 1A illustrates some-one whose alpha waves are strongly providing a bridge to the subconscious mind for the transmission of subconscious material up to the conscious beta waves. The problem is that there are no theta waves to provide this subconscious content.

My next step, in a case such as this, would be to help the client access theta. This approach almost always works, but Figure 1B shows how in some cases, the mind tries another strategy to avoid conscious contact with theta content. As soon as theta appears, the alpha that had been so strong only moments ago disappears. The material is there, but the bridge is gone! I have seen people shift back and forth between these two patterns repeatedly during a session, activating first alpha then theta, but never the combination necessary to sustain access to the content.

My solution to this problem is to work with the content of the block itself by means of a two-pronged approach that utilizes both state and content working together to break through the block.

In the face of this obviously elusive, repressed material, my client and I together must also face the decision whether or not it is wise to continue working on retrieving an experience that could end up being quite traumatic. One client I will refer to as S.P. had this exact pattern for many sessions. She knew that there was something in her past that was preventing her from going deep into her subconscious

1A **ELUSIVE CONTENT** **1B**

with full recall. After long discussion, S.P. decided she really wanted to pursue full awareness of whatever had happened to her. Her memories returned only sporadically and only in kinesthetic imagery and taste, never as visual or auditory.

A grisly but incomplete story emerged from the pieces S.P. retrieved. I encouraged her to move slowly and look for the imagery only when she felt safe to experience it. I also explained to her that she did not need to know the exact details of the childhood occurrences in order to heal. Finally, after working on this herself and with me for over a year, she had enough information to be able to let go of the need for more. She was able to achieve a resolution of the issues she had been dealing with and move on in her life.

Although S.P.'s brainwave pattern still showed a small amount of elusive theta when she was working in that area, she was happy to let it go. She then felt ready and able to work on building her theta waves with a more spiritual focus. Her continuing meditation practice concentrated more on connecting with a higher power and finding the right path on her life's journey.

The decisions about how deep to dig and when to let go are very personal ones. I never force a client to look at buried material. Yet when they are really keen on accessing it, and when it is getting in the way of their continued development, I will definitely help them work through whatever is concealed to a place of satisfactory resolution. And that place differs with each person. Again, if you feel unable to deal with your subconscious content on your own, I strongly suggest you enlist the support of a professional. Danger signals to watch out for are feeling overwhelmed, afraid, or depressed, or having any negative emotion connected with your endeavors that will not go away or threatens your stability in any way.

Most people, however, can feel confident and safe in pursuing an exploration of suppressed content through meditation and brainwave mastery. There are a number of ways to approach this.

USING YOUR STATE TO ACCESS YOUR CONTENT

Direct biofeedback comes in handy here. From the pattern of alpha/no theta, I take someone down into theta waves with guided meditation. *As soon as* the alpha starts to disappear, I give them that feedback and ask them to bring it back. Most people, with practice, can learn to sustain alpha while accessing theta. For some people, however, the theta is extremely resistant to being present with alpha. Usually in these cases the theta is very strong by itself, indicating a lot of repressed material.

Here I take the reverse approach, forgetting about the alpha initially and taking the individual into a deep and strong theta state. Then I carefully access a little bit of alpha. Letting a few images return at a time, I begin to drain the subconscious of its material. Teaching the meditator to sustain theta while developing alpha can be a more effective way of retrieving suppressed memories, but it is harder to accomplish this on your own without guidance.

''HOW CAN I DO THIS BY MYSELF?''

There are steps you can take to unlock blocks to subconscious material. The following is a form of meditation and internal dialoguing to help you move through the blocks caused by repressed material. With this meditation, allow yourself plenty of time and disturbance-free space to practice. As usual, read all of the instructions through before you begin and familiarize yourself with the steps you are going to take. Precede the meditation with a relaxation, but do not relax so deeply that you lose conscious awareness. You may also wish to have some gentle meditation music playing *very softly—barely audible, so as not to be distracting—*in the background.

THE MANY PARTS OF OURSELVES

There is one basic principle necessary for practicing the following exercises. *Each of us is made up of many parts. All of these different aspects of ourselves work together or separately to create the complex beings that we are. There is a basic ecology within this system that assumes that each part is ultimately working for the good of the individual.* Every part, in other words, has a positive purpose. It often doesn't seem that way, because somewhere along the line this part of ourselves has been misguided or misdirected to do something that in reality is now damaging. But the original role the part was intended to play was *for the benefit of the individual,* no matter how detrimental that role later becomes.

EXPLORING THE BLOCK

1. Before you begin, consider what you want to look for inside. For purposes of this meditation, we will stay very generic and focus on *the block itself*—not what is behind the block. That will come later.

(***)

2. Still your mind and relax your body. Let go of any preconceptions and expectations of any particular solution or outcome.

(***)

3. Create a framework of imagery for yourself to meditate within in order to access alpha waves.

(*** — or longer if needed)

4. Take your meditation deeper—go deeply *inside yourself* to create theta waves.

(*** — or longer)

5. Slowly begin to get in touch with the part of you that created the block. It's not important to have a crystal-clear image of that part. The more clearly you can bring it into focus, the easier it will be to communicate with it later on. At this stage of the meditation, however, do the best you can. All you really need is to connect with the sense of knowing that the part exists.

(***)

6. Find some way to manifest this part as clearly as you can—some way to connect with it. This may take the form of a symbol, image, color, body sensation, or other sensual representation for the *part of you that created the block.*

(***)

7. Begin to internally dialogue with it, so that you can make further connection and get information. You can ask it if it is willing to communicate with you. If the answer is yes, proceed. If the answer is no, ask it what it would take to allow you to communicate. If there is no answer at all, proceed as if the answer had been yes. You may hear the answers in words. You may experience them as images or sense them through body sensations, or it may just be a kind of a knowing. Remember to trust what you are getting.

(***)

8. Dialogue with this part of yourself. You might begin by asking, "How are you feeling?" Let this part express itself to you in whatever form it is able. You might want to ask, "How long have you been there?" or "What do you need?"

(pause thirty seconds to one minute)

Continue to dialogue with this subconscious part of yourself to get closer to it, to understand why it is doing what it is doing. You might ask, "What role do you play in my life?" or "What is your positive purpose for me?" or "How do you serve me?" You may also find your own specific questions. Take some time on your own to continue your exploration.

(pause for at least two minutes)

9. Express your appreciation to this part. Because it has been working *for* you it deserves recognition and appreciation, even if it has ultimately caused you harm.

(★★★)

This step is hard for some people. They may have so much anger at a part that has been blocking their progress that it is hard to understand that the reason for the block has been self-protection or survival. Have faith here that your subconscious, although misguided, has your best intentions at heart. Giving it appreciation *for those intentions* does not mean condoning continuing use of its methods.

(★★)

10. Now explain to this part how *you* are feeling. *Even if your conscious mind realizes fully how this part has been blocking you or is detrimental to you, DON'T EXPECT YOUR SUBCONSCIOUS MIND TO UNDERSTAND without careful explanation.* Don't blame. Use all of the communication skills you would use with another human being. Tell the part how its behavior has been affecting you (what having the block has meant to you and the difficulties it has created).

(★★★)

11. Consider what you would like to have happen now. Look at the original reason for the block. Do you still need it, or has that need changed? What do you actually want your subconscious to do?

(★★★)

12. Tell your subconscious what you want. Ask it to help you get what you need for yourself right now. If necessary, remind it that it is there to benefit you and what it has been doing in the past no longer serves you now.

(***)

13. Negotiate. Actually bargain with your subconscious. You can continue to use questions to help you with this communication. If you still need the *role* that this part plays to be active within you (for example, for protection or survival), ask if it can fulfill this role in a different way —one that would no longer require you to experience the negative side effect of the block. If you no longer need this service, tell it that, and ask if it would be willing to stop doing it. Useful questions also might be "What would it take for you to stop blocking me?" or "What needs to happen to make a change for the better?"

(***)

14. Come to a deal with your subconscious. Perhaps you need to put it in a time frame or within some other context in your life, e.g., "If I do this . . . then you will allow this block to disappear."

(***)

15. Thank this part for its willingness to cooperate.

(**)

16. Prepare for closure. Ask your subconscious part if there is anything else that it wants to communicate with you right now. Tell it any final things you would like to tell it before you complete your meditation. You may want to make plans to check back with the subconscious material at a specified time in the future to see if this reprogramming is working or if there are any unforeseen problems that have arisen.

(***)

17. Complete your meditation and reawaken, taking several deep rapid breaths and stretching fully.

(***)

18. Ground the meditation by putting your experience into words—writing it down or talking about it.

(pause for as long as you need)

If you cannot find a positive purpose to begin with, keep looking. And keep asking questions inside. Track it back as far as you need to in order to find the original positive purpose.

Example: A woman wants to find the positive purpose for the internal part of herself that is creating her intense emotional pain.

She asks the pain, "What is your role in my life? What positive purpose do you have?" The pain replies, "To punish you!"

Now this doesn't seem like a *positive* purpose, so she doesn't stop here. She asks, "What is the positive purpose of punishing me? How does that serve me?" The pain replies, "To make you feel bad about yourself."

Again no positive purpose. So she asks, "How does feeling bad about myself serve me?" The pain replies, "To keep you in line." Now she is starting to make progress, though she doesn't yet know where this is leading her.

Following this track, she asks, "How does keeping me in line serve me?" The pain answers, "It keeps you from making mistakes."

Feeling this is still not a positive enough answer, she asks, "How does keeping me from making mistakes serve me?" She knows she has finally gotten to the root when her pain answers, "It protects you from danger."

After tracking her pain back this far, she realizes that what she has been feeling is actually a mechanism she set up in her subconscious to protect herself from the danger of making mistakes. Though this is no longer a logical, effective setup, it might well have been when it was initially put into place. Dialoguing with her pain further, she discovered that it had been created as a defense against her somewhat abusive mother, who threatened her when she made mistakes. To her young mind, it became imperative for her self-protection that she not make mistakes, or at least make as few mistakes as possible. So her young subconscious set up this system.

If she felt enough pain, she would remember to stay in line, so that she wouldn't make a mistake and thereby be in danger of her mother's disapproval and possible abuse. Our subconscious minds do remarkable things in the name of our own protection and survival.

THE FIRE

A middle-aged construction worker came to me with the complaint that he could not visualize, nor had he ever remembered a dream. A brainwave profile showed the pattern of suppressed content. After only one session of working on it, the memory returned to him of a severe fire he had been in as a very young child.

The images that he saw during that experience were so horrific that he had made a subconscious decision never again to allow himself to visualize. The memory of the fire was repressed deep within his theta waves, and his brainwave pattern took care to never put him in a state of consciousness that might possibly allow the memory, and therefore the terrifying and gory images, to surface. After only a few sessions of working with this and allowing the memories to return and be expressed, he began to remember his dreams and was able to visualize during meditation.

Sometimes becoming conscious of the contents behind blocks is not so easy. The blocks are more complex and the stakes in remembering them are higher. It is advisable to go slowly and respectfully into subconscious content that has been intentionally buried. If you think you are going to have trouble dealing with the conscious awareness of this material when it is exhumed, engage the services of a counselor or therapist who understands the workings of the subconscious mind.

BEHIND THE WALL

One woman came to me in a lot of pain and fear. She had been unable to take a shower since she was eleven years old. Whenever she tried to close the shower curtain, she was gripped with fear. She also had a memory gap of several years surrounding this period.

Using the kind of internal dialogue described above, I asked her to get in touch with the block. The image for the block was a high, thick gray wall that she could not go around or over. Through several sessions of delicate work, she got the wall to agree to let her see through it. We thought we would finally come to what was behind the memory block. When she got through the wall all she saw was a mountain—another block. We had to do considerably more work to get through this block. Finally, the memories started *very slowly* to trickle in. She remembered her brothers abusing her every time she took a shower.

This kind of inner dialogue can be adapted for the exploration of almost any kind of subconscious material. For example, if you feel disorganized or lazy, find a symbol or image for the part creating your disorganization or laziness and find out what positive purpose it serves. The steps are the same. The questions you ask are the same. You simply have a different central theme.

You could do the same with fear of success, self-sabotage, or any emotion, such as desperation, anxiety, guilt, anger, or panic. Any time there is *any feeling, need, attitude, behavior, or difficulty inside that you don't understand, you can look to your subconscious for information and guidance. You can make changes in your subconscious wiring to be healthier and happier and have a more balanced, stable, and clear psyche.*

AN EXPERIENCE OF TRANSCENDENCE

A client I shall call C.R. is a thirty-two-year-old therapist who has done many years of personal growth work and is familiar with meditation and healing states. Her initial brainwave profile revealed a strong tendency toward an awakened mind pattern in meditation, with an overabundance of theta and delta cut off from her conscious mind by an inconsistent alpha bridge. She wanted to "get the subconscious out and strong in its own right."

During the second and third sessions, we worked on stabilizing her alpha and helping her maintain it while accessing the contents of the subconscious. The block she found was a brick wall protecting her from an entanglement of abuse and spirituality. Through childhood abuse, she had somehow intertwined the two, which left her feeling as an adult that being abused had some spiritual meaning or necessity. At this point we realized that we needed to allow her strong spiritual nature to manifest more fully, in order to heal the abuse.

She came to the fourth session with two major issues. In therapy she had contacted an image of herself at ten years old, a radiant young girl with long, blond hair who said, "Stop interfering. You never trust me. Leave me alone. I need you to trust my process and stop trying to control it." C.R. felt this girl was the key to something that she had been looking for for a very long time. She also wanted to explore the issues in her life surrounding friendship. C.R. often felt lonely and in pain because she experienced her friends as not being as available as she would like them to be. She wanted to examine why feeling "connected, seen, and contained" by another person was so necessary for her sense of well-being. Her hope was to find a better way to satisfy those needs without placing so many demands on others and consequently being hurt by their rejections.

As I took her into meditation, she went quickly toward an awakened mind pattern with her characteristically strong subconscious and unconscious. She then contacted the ten-year-old girl, who began a process of transformation. As C.R. described the experience, "She is becoming some kind of a divine being, exuding strength and light. She is fierce, powerful, beautiful, and gentle all at the same time . . . I am moved, deeply touched. I can feel a place inside open where I have held on for a long time."

I asked her whether this being could provide the support and connection she was unsuccessfully looking for from friends. C.R. struggled with embarrassment, denial, and fear before she revealed the words this radiant being said to her. "I am essence, I am the light, I am eternal, I am always here whether you recognize me or not." She reported feeling surrounded by knowing, by a sense of "beingness that requires no action or thought," and by peace.

At this point she began to wonder why she hadn't been able to experience this state before. Had she been afraid that allowing it into her consciousness would take away her need for *any* friendships? Eventually she realized that she had been worried that she would become a hermit if she got in touch with the transcendent part of herself. Because of this, she had inadvertently expected her friends and relationships to provide this state for her when, in reality, she could find it only within herself.

Since this session C.R.'s life has gone through both external and internal changes. She put her relationship on hold and traveled for many months. She wrote, "A great deal of reorganizing has taken place of what I think, how I view the world, how I relate to others. I have talked about the session with few, since words seem so insufficient for the task and because I feel protective of what is growing inside of me as a result of the experience I had. One of the most important shifts that resulted from the session is realizing that the abuse I experienced as a child did not damage my essence—that at the core I am whole. A wound that has festered for many years can now begin to heal . . . can I finally relax and experience the internal rest that I have been searching for all my life?"

THE MATERIAL OF THE MIND

The high-performance mind is created by the combination of the state of consciousness and the content of consciousness. As we have seen, the optimum state of consciousness to support the high-performance mind is the brainwave combination of beta, alpha, theta, and delta called the awakened mind. Judicious use of the content held within those brainwaves will give us the self-knowledge we are seeking. This chapter has looked at the development and manipulation of the material of the mind through meditation. The next chapters will explore the high-performance mind in healing and creativity.

HEALING

To heal means to make whole, to make well, to make sound. When we look at healing, we can look at it in two aspects—self-healing and healing others. Healing ourselves means to use every resource available to us to make ourselves physically, emotionally, mentally, and spiritually whole. Healing others means to use our resources—sometimes those same resources that we use on ourselves—to help other people become whole.

SELF-HEALING

Self-healing does not mean that you must heal alone—that you have to do it all by yourself. Self-healing means that you take the responsibility for your own health yourself, rather than give your choices and power away to another person. Self-healing might mean that you research and try the best possible treatments and medications available from as many sources as possible, making informed and deeply considered choices about the ramifications of each method of treatment. Your best method of healing might mean not only to choose the path of surgery, but to seek out the best surgeon in the country for the task. Or healing might mean to delay or forgo surgery and instead choose a combination of herbal remedies, energetic medicine, spiritual healing, and meditation. You might choose to supplement this with allopathic medication if the situation warrants. *Self-healing means that* YOU *make these decisions and* YOU *continually monitor the process of your progress and alter your choices accordingly.*

How are we able to make those choices? By study, research, meditation, and faith in the guidance of a power greater than ourselves. In order to be in the best possible *mental state* for taking this responsibility, we can develop a balanced, optimum brainwave pattern. By developing self-mastery of *any* biological function, we increase our ability to master other biological functions. In taking responsibility over our own body's functions, we can learn to lower our heart rate and blood pressure, raise our temperature, redirect our blood flow, improve our digestion. The state of consciousness that we are in while learning to do these things is vital to the process of self-healing.

The brainwaves state that is optimum for self-healing is, once again, usually the combination of beta, alpha, theta, and delta that we know as the awakened mind pattern. While experiencing this pattern, we can utilize all aspects and levels of consciousness to benefit our health. This is especially true for self-healing meditations that involve imagery of any kind. The theta brainwaves allow the imagery to come from and return to the subconscious, deep levels of the individual, where it can continue to create effective input. The alpha brainwaves provide the bridge from the subconscious mind to the conscious mind so that the individual can maintain an open flow of sensory images and information important to the healing process. Beta brainwaves provide the conscious mind access to the subconscious mind through the alpha bridge. The individual is able to communicate his conscious thoughts, ideas, and needs to his own inner being. Likewise, he is able to understand, on a conscious level, the deep inner needs, feelings, and experiences that are stored or hidden away inside him.

PSYCHOPHYSIOLOGICAL RELAXATION
AS SELF-HEALING

Though the optimum brainwave state for self-healing is usually the awakened mind, deep mental and physical relaxation is also useful. This type of meditation does not require the meditator to maintain conscious awareness and alertness throughout the meditation. Instead of using alpha waves to create a bridge to the subconscious, the subject dives down into the mind as deeply as possible. If we wish to position this state on the graph on page 44, we would place it in the extreme bottom left-hand corner.

Note that even if the meditator is producing only delta, the physical relaxation is well below the normal sleep state, which indicates that she is in a form of deep trance state. With this depth of physical and mental relaxation, the individual's body and mind are put into a restorative, regenerative state, a very deep form of rest. An hour of this kind of rest can be the equivalent of several hours of sleep. Imagine if you meditate two hours a day in this deep restorative state *and* get your normal six to eight hours of sleep a night—your body would have so much more time to heal.

This method of self-healing is especially useful for serious systemic illnesses, when the body is in need of deep rest. Max Cade was able to help a student of his to heal herself from cancer by inducing her into a state of deep psychophysiological relaxation for long periods of time daily. The depth of the relaxation and the length of time spent in that state eventually helped her cancer to go into remission.

If you practice this type of meditation when you are not ill or in need of extra rest, you may find yourself not needing as much sleep at night. In fact I have seen this form of meditation used when an individual wants to do without sleep for a period of time. I knew of a woman, again a student of Cade's, who needed a great deal of study time to pass her college exams. She put herself into a state of deep psychophysiological relaxation and meditated an hour a day at eight o'clock in the morning and an hour a day at eight o'clock in the evening for several weeks without sleep. She passed her exams! I can't, however, attest to the state of her health at the end of this process, and it is not something that I would recommend. I believe the body needs downtime, not just deep rest.

I believe that deep psychophysiological relaxation works best in conjunction with other methods of self-healing meditation. The contrasting methods cannot easily be practiced in the same meditation, because they require very different brainwave states. They can, however, be alternated if that is appropriate.

A far more common form of self-healing meditation is a participatory, active style utilizing sensual imagery. Two major approaches, imaging a change occurring from ill to well and imaging yourself as already well, are outlined below.

THE BRAINWAVES OF
SELF-HEALING IMAGERY

If you are seeking to develop an awakened mind brainwave pattern, following specific styles of self-healing meditations provides an excellent brainwave-training practice. Alternately, if you want to have the most effective self-healing meditation, practicing it while in an awakened mind brainwave state will be the most effective. This is a prime example of content helping state and state helping content *at the same time.*

Self-healing meditations are practiced in the traditional meditation brainwave state of alpha, theta, and, optimally, delta, with the addition of beta. This is a *working meditation.* You need the beta in order to add the *conscious* content of healing. The sensualization exercises you have used to strengthen your alpha find an important application here, and the depth of theta is vital to implanting the suggestions and images of self-healing into the subconscious mind, where they can begin immediate and effective work.

IMAGING A CHANGE FROM
ILL TO WELL

Carl and Stephanie Simonton made the techniques of visualizing self-healing famous in their work with cancer patients. In their seminal book, *Getting Well Again* (J. P. Tarcher, 1978), the Simontons described how they used mental imagery for self-healing. An abbreviated example of the form this mental imagery might take is "See the cancer as raw hamburger meat and picture the white blood cells as big white dogs gobbling up the meat."

While these techniques had a dramatic effect on the nature of self-healing and on many medical practitioners' and patients' attitudes, the authors were still talking in terms of "seeing" and "picturing." The visual image was king. This form of self-healing was also effectively applied to other types of illness, as long as the individual could see imagery.

As we noted earlier, people whose primary access sense is not visual or people who actually cannot see pictures in their minds might feel that this form of self-healing is unavailable to them. Not so. You could just as easily *hear* the illness as a cacophony and the healing as beautiful strains of violin music coming in, overpowering and replacing the noise. The cancer could be a sour taste and the cure the taste of sweet nectar. The illness could be a rough abrasive sensation, while the cure feels smooth and velvety. The sickness could be a putrid, rotten smell and the cure the scent of flowers or perfumes. Or perhaps the disease could be jerky staccato movements, while the cure is smooth, flowing, and fluid motion. The optimum mental image for self-healing would include a picture, a sound, a smell, a taste, a texture, and a kinesthetic sensation combined into one.

Sensualization techniques become vitally important when we are using them not just to develop alpha waves but also to plant the strongest possible positive images of healing throughout our unconscious, subconscious, and conscious minds.

IMAGING YOURSELF AS
ALREADY WELL

The main complaint about the illness-to-wellness type of imagery model is the following: What happens if the desire for wellness is not strong enough? if the fear of the illness or its power has a stranglehold on the meditator? if the clarity, depth, and sensualization of the illness is *stronger* than the clarity, depth, and sensualization of the cure?

If the meditator puts more focus on the illness than on the cure, the condition of illness could be enhanced by the very mechanism you are hoping to use to cure it.

A second kind of self-healing imagery meditation, therefore, is to sensualize yourself as already well. Imagine what you are like *after a cure has taken place*. Using all of your sensualization ability, create a mental image of health—without thought of or focus on the original illness.

Both of these methods of meditation can and do work for people. When I teach self-healing meditations in a group, I like to suggest that people use both methods. If they use the first method, which is a very powerful form of self-healing, they should close the meditation by using the second. This way, if they had a difficult time sensualizing the cure or were in danger of giving added energy to the illness, they can use the second meditation as a kind of fail-safe healing image. I suggest that you always close your healing imagery with positive sensualization of what you will be like after the illness is completely gone.

PREPARATION FOR THE MEDITATION

Body-image meditations can help you develop the experience of shifting your body in some way. Practice these as you did the sensualization exercises in Chapter Four. Begin by allowing yourself to become fully relaxed and entering into your meditative state.

1. Imagine your hands becoming hot. You can use any method of sensualization that you like to do this. Experience yourself washing dishes in hot, soapy water or holding your hands out in front of a hot fire, a blazing log fire. Be careful not to burn them. Really *feel* them warming. You may even be able to feel the blood flowing into your fingers—almost a pulsating, throbbing kind of feeling as your hands and fingers become warmer and warmer.

2. Then feel your forehead getting cold, very cold. You could imagine yourself walking out in the snow on a cold winter's day, all bundled up except for your forehead, the wet snow hitting your forehead, making it cold. Or put an ice cube to your forehead, but be careful not to lose the heat in your hands.

3. Try to imagine your forehead cold and your hands hot at the same time: your hands hot and your fingers throbbing, actually swelling a little as the blood enters them; and your forehead cool, a cold spot right in the center of your forehead.

4. Now allow those images to fade, and return to an awareness of your whole body.

5. Begin to feel your body becoming heavy—heavier and heavier, doubling your weight, tripling your weight—so heavy that you couldn't lift your arm if you tried, so heavy that you couldn't lift your eyelids. Heavy.

6. And now become lighter again, gradually lighter and lighter—so light that you feel almost as if you are going to float away, so light that only the weight of your clothes is keeping you in your seat. Light.

7. Slowly return back to normal, the weight you are used to.

8. Now begin to experience yourself shrinking—smaller and smaller, half of your size, a quarter of your size—shrinking until you are very, very tiny. You could fit into a kitchen matchbox. In fact, you may want to try that now, finding an open matchbox with only a few matches left

inside. You hoist yourself up and crawl inside. What is it like climbing over the matches?

9. Then you are growing again, growing to your normal size and beyond, larger and larger—eight feet, ten feet, and larger still, until you are double, then triple your normal size. Now go sit in a park outside with your head above the trees and experience what it is like to look down at the houses and people below you.

10. Now shrink smaller again, and smaller, back to your normal size. Again be aware of your whole body—your *whole* body.

11. Now experience your meditative self, the feeling that you have when you are in your deepest state of meditation, trance, or relaxation. Experience how your body feels and what you experience when you are in deep meditation, and enter that state.

SELF-HEALING MEDITATION 1

Here are the steps for a self-healing meditation that uses imagery to create change. As always, use your strongest sensualization technique, calling upon as many senses as possible to experience the images that you are creating inside.

1. Consider the problem that you wish to work on and then put it in the back of your mind.

2. Relax your body and still your mind, going into a deep meditation. Maintain enough alpha to remain alert and aware of what you are experiencing.

3. Consider the problem that you wish to heal and create an image to represent it. Your image can be a picture, a sound, a smell, a texture, a taste, a sensation, or any combination of these. It can represent the problem realistically or symbolically.

4. Find an image that will heal the problem. This is an active part of the meditation. Imagine the healing image coming into your field of awareness and beginning in some way to transform and heal the image of the illness or the problem.

5. Sensualize it working. Spend several minutes actually experiencing the transformation of the images.

6. If you feel finished with that image but want to continue in this phase of the meditation, you can try another image for the problem and another image that will heal it.

7. Leave it working. Create a setup inside yourself to leave the transformation in progress, so that when you finish the meditation, the healing will continue throughout the day, or even throughout the week.

8. Imagine yourself as already healed. Sensualize what you would feel like if the problem or illness *simply did not exist.* Using all of your senses, experience yourself as well.

9. Look into the future and imagine two or three situations where you would normally have been affected or limited by the illness, and experience yourself as being completely well and fully functioning.

10. Lock this image inside yourself. Find a way of anchoring it in, so that it will continue to be present inside you even after your meditation is complete.

11. Find a closure for your meditation. Complete any unfinished business that you have. Allow yourself to return from your meditation feeling rested and relaxed, yet fully alert and present. Take several deep, strong, arousing breaths. Let yourself stretch fully and open your eyes.

12. Ground your meditation by writing, talking about, or drawing your experience.

SECONDARY GAIN

Before we go on to the next type of self-healing meditation, we need to address a very important element of the healing process that may be relevant to you. If your illness has a hidden benefit or a positive reason, that is called *secondary gain*. If you ask yourself, "How is this illness serving me?" you may find that there is a secondary gain involved.

Examples of secondary gain range from the common experience of getting sick on the day of a big exam in school (the gain is not having to take the exam) to receiving the love and attention you never got from your spouse until after you contracted cancer (the gain is love and attention).

Secondary gain can be more complex and less easily identifiable than those two easy examples. I had a client with extreme allergies. Through our theta work we determined that she was actually allergic to *men* rather than substances. The secondary gain of her allergic reactions was to protect her from developing intimate relationships with the opposite sex, something she was desperately afraid of. We didn't need to work on healing the allergies at all. Instead we needed to work on healing her relationship issues with men. When that inner processing was completed, her allergies faded quickly away.

THE NEW AGE DILEMMA

We need to be careful here, lest we go overboard with self-responsibility. A common question among those in the New Age self-healing circles is "Why did I cause that?" or "Why do I need that illness?" We're walking a fine edge here. On one hand, *everything* that happens to us is somehow our responsibility, ranging from "Why did I choose *these* parents?" on up. We can get into a potential danger zone if we try to read hidden meaning and responsibility into every aspect of our health and well-being.

I believe in the theory of germs! It is quite possible to catch an illness because we are standing next to a person who transmits it to us. On the other hand, the sincerely dedicated might ask, "Why did I need that germ at that particular time?" One answer might be, "I was working too hard and got run down, therefore I was more susceptible to the germ than I would normally have been." In the same vein one might ask, "Do I perhaps need this illness as a reason to take a few days off from work so I can get some rest? Would it therefore be possible to take those days off from work *before* I get ill and thereby prevent the illness?" If we can find the hidden agenda or positive purpose in the illness and find another way of solving the problem without getting sick, then the illness might be avoided altogether.

Along the same line, if we simply try to heal the problem without looking for the root cause, we are successful only for a time. Soon a new health problem pops up out of the same root cause. We might stop smoking cigarettes only to develop an eating disorder. We might heal the nausea but develop an ulcer, until we find out what it is in our life that we "can't stomach."

I work on a basic principle of ecology. I believe that all parts of a human being are working in some way for the good of the whole, no matter how misguided they may be.

The root cause of a problem may initially appear to be negative— often a form of punishment. If you do this work, it is essential that you track the negative aspects of your problem back to their original positive purpose. Let's look at a few examples of this process to get a deeper understanding of how it works.

1. A man has a severe case of eczema. He goes into deep meditation and contacts the part of himself that is responsible for causing the disfiguring skin condition. He sees an image of an ugly, scaly face and asks it what the purpose of the eczema is. His inner part says, "To make you ugly." We cannot stop here. This is not a positive purpose. He asks, "What is the positive purpose in me being ugly?" "To keep you from going out with women!" "What is the positive purpose of keeping me from going out with women?" "To keep you from getting married"—still not a positive purpose. "What is the positive purpose of keeping me from getting married?" *"To protect you!"* At last we have found our *positive* purpose.

With this knowledge, he can then decide for himself whether or not he still feels the need for this kind of protection. His parents had a painfully broken marriage, and his inner psyche was simply doing its best to protect him from that experience. He may be able to work through his feelings about marriage and realize he no longer needs protection from it. Or he may still feel that he doesn't want to get married, but with that knowledge now available to him on a conscious level, he might choose more appropriate ways of maintaining his unmarried status. He no longer needs the protection of the eczema.

2. Even the most distressing of all inner parts has a positive purpose if you look to the ecology of the human being. A woman is on the verge of suicide. She has a part deep inside that wants to make her kill herself. How can we possibly find a positive purpose in that? We track it back. What is the positive purpose of death? Escape. What is the positive purpose of escape? No pain. The suicidal part simply wants to relieve her of her immense pain. We then need to look at the cause of the pain—the part inside that is creating it—its positive purpose, and so on. This work can become long and involved. However, it can provide an invaluable psychological unknotting of the subconscious and a clearing of long-stored attitudes, belief systems, and ways of being in the world.

It is important that this work be done on a theta-brainwave level. For permanent changes to take place within the subconscious, the subconscious has to be accessed. Just thinking about the problem and talking about it in beta waves is not going to cause an effective or permanent healing to take place.

The following meditation deals with the issue of secondary gain.

SELF-HEALING MEDITATION 2

BE AWARE OF THE ISSUE OR PROBLEM THAT NEEDS TRANSFORMATION.

Consider what it is that you want to have healed. Put it in the back of your mind and leave it there. Then allow your mind to clear.

STILL YOUR MIND AND RELAX YOUR BODY.

Remember what it feels like to be in your state of meditation and allow yourself to return there.

Focus on your breathing and gently let yourself breathe easily and deeply, using your hand as a lever to slow your breathing down for a minute or two if you wish.

Remember to relax your tongue. If you feel any pulling on the back of your tongue it means that you are talking to yourself. If you relax your tongue completely, you can't talk to yourself, and it will be much more difficult to think. Anytime an unwanted thought comes through, relax your tongue as you exhale and breathe the thought away. Breathe relaxation into your mind when you inhale, and breathe away thoughts when you exhale . . .

Relax your body. Check to make sure that the *whole* of your body is relaxed. Gently scan your body to make sure you are not holding tension anywhere. Shine a light through it. Notice any areas that are darker than others. Breathe relaxation into those parts of you when you inhale, and breathe away the tension when you exhale. Continue relaxing your body until you can see the light throughout.

CREATE A HEALING ENVIRONMENT.
In your mind create an environment that is healing for you. It can be indoors or outdoors. Walk around in this environment. Notice the colors . . . the shapes . . . the forms. Be aware of the textures, the atmosphere, the temperature . . . the sounds, the smells, even the tastes. Explore this environment using all of your senses. Make any changes that you want to make, adding anything that you might need for your health and well-being. Create a space that is secure, serene, and safe.

MEDITATION
Find a comfortable place within this environment to sit down, and allow yourself to go into a very deep meditation.

Experience a sense of falling . . . falling deeply into relaxation, into warmth . . . into yourself.

THE PART THAT NEEDS HEALING
From this place of depth inside yourself, begin to get in touch with the part of yourself that needs healing, transformation, or change. It may be the issue that you originally thought about before the meditation began, or you may find that a deeper or more meaningful issue comes forward, ready to be healed.

MANIFESTATION OF THE PART
Find some way to allow this part to manifest for you. It could be through a symbol, an image, a body sensation, a feeling, a sound, or even a voice in your head. Or it could simply be a sense of knowing that the part is present.

BEGIN TO DIALOGUE.

Ask if it is willing to communicate with you. If it is uncertain, ask what needs to happen to make it more willing. (What follows is a series of questions that you might like to ask this part of you. Each meditation and healing experience is different, so use these only as a guide. You may add, delete, or adapt in any way that is helpful for your process.)

—How is that part of you feeling? If it is feeling at all threatened by this experience, you might like to reassure it that what you are in the process of doing is for the benefit of your being as a whole . . . that you are doing this to heal and to help, not to hurt.

—How long has it been there? Is this a new issue or a very old one? This may lead you to ask, How did it get there?

—What does it need or want from you?

—What role or purpose does it play in your life? What is its positive purpose for you? If the purpose seems negative or obscure, track it back until you can discover how this issue or problem serves you.

—Is this part playing a role you still need in your life? If so, how can this role be fulfilled in a way that is more beneficial to you?

—If not, how can you transform this part? If you no longer want this service performed for you, gently but decisively inform the part that its role is no longer needed in that way.

—What needs to happen if this part is to work in a way that is healthy and beneficial to you in the present?

—Is this part willing to let this change occur? (If not, what needs to happen to allow it to become willing? Remind it that you are doing this for the benefit of the whole of you right now, and that your needs have changed, probably dramatically, since this part was set up to fulfill its particular role.)

Sensualize. Use as many of your senses as possible to imagine the change taking place.

Give your appreciation. Acknowledge that part for all of the hard work that it has done for you. Even if it has fulfilled a function that you disagree with *now,* offer your appreciation for the fact that, inside yourself, this part of you thought that it was working for your benefit.

Allow the part to respond to you. Ask it if there is anything else that it wants to say to you.

CHECK TO SEE IF ANY OTHER ISSUES HAVE BEEN STIMULATED.

Are there any other parts inside you that want to say something to you? This inner change may have stimulated *other* aspects of yourself that are happy, sad, frightened, disturbed, or pleased with the transformation. Give them an opportunity to communicate with you.

If necessary, begin your dialogue again with the new part that wishes to communicate. Follow it through to completion.

CLOSURE

When you are ready to complete your meditation, spend a few minutes inside preparing to close, giving your appreciation to your inner parts for their willingness to participate in your healing. You may wish to make an inner agreement to check back inside yourself at some future time to continue this process or to allow yourself further opportunity to meditate. A closure is not necessarily an ending.

BRING IT TO BETA.

Crystalize your experience in a few key images, words, or phrases. Bring those out with you when you awaken from your meditation. As always, verbalizing, talking into a tape recorder, telling a friend, or writing down what happened will fix the experience firmly in your conscious beta mind so that it will remain present for you and not slip back into your subconscious.

Make sure that you stretch, breathe deeply, and arouse properly before you leave your meditation space.

You may find that by doing this meditation you initiate a process in yourself, and that it would be useful for you to maintain and monitor that process over time. Self-healing is not a one-shot event. It requires dedication, patience, and belief in your own innate recovery abilities. You can try these meditations as often as once a day for two weeks or more, and observe the changes that have taken place from day to day and week to week. Especially if you contact a weak, new, or previously hidden part of yourself, it might need encouragement and attention on a frequent basis to be allowed to develop to its full potential or to be nurtured into a state of well-being.

The variety and intricacy of the imagery that I see surface when I am doing this work with individual clients and groups of students never ceases to intrigue me. Frequently I cannot second-guess the resolutions that are to be negotiated, because they are so deeply private, involving so many factors from a person's past and present. Often when people share their meditations with me it feels like going down the rabbit hole with Alice in Wonderland. I greatly enjoy this aspect of my work. In order to be the most helpful, I need to stay absolutely present with their processes and not project any possible solution before they come to it themselves. Only if an individual gets stuck and cannot proceed do I intervene with suggestions. My role in these meditations is primarily asking question after question to help the inner healing process reveal itself.

HEALING THE LIVER—E.G.

The following is an account of the experience one of my students had while participating in the above meditation during a group at Esalen Institute. His images are particularly vivid. Although it is not necessary to have such lucid detail in order for the healing to be effective, clearer imagery will promote more effective healing, even when the content is very generalized or simplistic.

The subject I had originally chosen to deal with in my healing meditation absolutely refused to materialize in any stable form. Instead, all my efforts were obscured by a vaguely defined collection of internal organs. The most solid of these was the liver. As far as I know, I have no liver problems or other internal problems; however I have been concerned about my diet—particularly the amount of junk food and alcohol I consume.

As I was considering this, the liver image, a rather shapeless dark blob, clarified and grew, pushing out all the other body parts. This was obviously my subject.

I approached the liver, and it resolved into a dense thicket of large bushes, almost a forest. I looked closer to see that the leaves were darkened with dust and soot. Some trash and other debris was scattered around in the branches and on the ground.

My first idea was to use an industrial steam jet to clean all this up. I created a huge steam hose and started blasting away at the darkest areas. When I did so the leaves and bark were stripped away with the dirt, leaving a bleached white skeleton of branches. Not a good idea!

Next I tried washing the leaves with soap and solvent chemicals, but the smell was bad and the leaves began to shrivel. They appeared to be healthy under the dirt. Then I started throwing buckets of water on to the leaves. This helped a little, but the job was so huge that I felt overwhelmed.

Then I had an inspiration—I brought dark clouds over and made it rain on my forest. As I stood and watched, I saw that some of the dust was washing off, but to my frustration, most of the dirt appeared resistant. Soon though, I noticed that new green sprouts were appearing among the dark leaves. Ferns and other plants were pushing up through the newly wet ground.

Letting the forest regenerate rather than trying to fix what was wrong seemed like the right thing to do. I spent the rest of the meditation walking around in the rain, encouraging green things to grow, making sure the soil was rich and moist, and picking up the occasional beer can or other piece of garbage.

Now I know I can go back to tend my forest at any time, to make sure that everything is growing properly and that the rain is still falling.

E.G. learned from his subconscious that his liver needed healing, and he found a way within himself to begin to do so through sensualization. A number of trials and errors were necessary before he found the images that worked for him. This is an excellent example of the value of letting the subconscious mind create the healing imagery, rather than trying to use beta—your own or someone else's —to impose a preordained process.

EXAMPLES OF SELF-HEALING

CESAREAN SECTION

In the middle of a cesarean section, a colleague and student of mine named Suzanne found herself in trouble. As the surgery proceeded, she became confused and frightened and felt herself slipping into shock as she began to experience the pain of what the doctors told her were "windows" in the anesthesia. Because the anesthesia was not affecting her properly, the anesthesiologist continued to increase the dosage in hopes of reducing her extreme pain.

The numbness began to spread upward into her chest, and she felt as if she could no longer breathe. She vaguely remembered hearing the doctors talk to each other about putting her on a respirator if her lungs stopped working. She desperately wanted to stay conscious for the birth of her child, but she felt herself slipping away.

At this point she realized that she did not have to be a victim of this situation. She remembered her training. She remembered me telling her that yogis could survive easily on one breath a minute. She remembers saying to herself, "If a yogi can breathe only once a minute, so can I." She stopped screaming and fighting for breath. She intentionally and consciously slowed her breathing down so that she could remain conscious on the small amount of oxygen that was still available to her from her failing lungs, and concentrated on eliminating the pain. Her image was one of moving through the pain, dispersing it, and dissipating it. She imagined that she was surrounded by the pain but that she was so centered that she was actually able to move between the "pain molecules" so that they did not touch her. She then became "detached and serene, waiting for the baby to come out." She proudly reports that she was present, conscious, and awake when her seven-pound baby boy took his first breath, gave his first cry, and was held to her breast. She fully attributes her experience to the ability to remember her meditation techniques and to realizing that *she* was in control, not the medication.

BLINDNESS

What better way to describe the benefits of self-healing meditation than to give you an example from my own life? I feel blessed to have survived and recovered from blindness. The path of the healer or the teacher is often carved by her own quest for healing. My personal journey has made a significant contribution to my understanding of this work and my drive to experience higher states of consciousness.

In 1975, while living in London, I began to have visual distortions, or blind spots, in the macula of my left eye. The eye specialist I was referred to identified it only as a "pimple on the retina." I was unconcerned, since there was no indication that this could be a serious or progressive condition. The visual distortion came and went several times until the summer of 1976, when it came and stayed. While on vacation in Greece, I watched the blind spot grow to obscure my central reading vision in that eye. Then, on a plane flight back to the United States for a visit, it hemorrhaged, and I knew I was in serious trouble.

What the doctors in England had been unable to diagnose, since the condition is so rare there, was readily diagnosed in Colorado as ocular histoplasmosis—a degeneration of the retina that I was told could and probably would occur in my right eye as well. On returning to England, I saw the top retinal specialist there, who confirmed the diagnosis and suggested that I learn braille. As my left eye continued to deteriorate, I was thrown into turmoil.

Because there was no medical treatment (no one in either country would laser something so close to the center of my vision), I was thrown back on my own resources to try to find a way to heal. Although I had learned about self-healing meditations from Max Cade, I hadn't had a real necessity to use them. I began meditating for approximately three hours a day in the styles I have described above to try to stop the bleeding in my eye. At the same time I began going to healers and researching who might be able to help me. My investigation sent me to Mary Simpkins, an eye healer in Eastbourne, England, who uses what she calls visible-ray therapy along with hands-on spiritual healing. She has developed an instrument that emits pulsating rays of light, which she alters in color and speed of pulsation according to the individual needs of her patients. Looking at these individual wavelengths within the visible ray spectrum can affect the molecular structure of the retina and help regenerate the degenerative area. The experience is rather like looking into a pair of extended binoculars with gently pulsating colors at the end of them for long periods of time.

Yes, she could help me. Unfortunately she had a seven-month waiting list. I had made a decision that blindness was not the path I wanted to follow in this lifetime, so I set about trying to maintain what sight I had while I waited to begin my treatment with Mary. I used many different forms of imagery, including imagining rays of white light sealing off the bleeding area in the retina, miniature sponges mopping up and removing the blood, and gentle abrasive sandpaper removing the scars. I practiced about an hour a day of imagery and about two hours a day of deep psychophysiological relaxation.

The process of healing was long and arduous. But at the same time it opened many doors to me. The kundalini experience that I refer to in the Appendix began, I believe, as a direct result of such intensive meditation—although that was by no means my focus or intention. My search for healing led me to experience the skills and powers of many of England's top healers. My heart and spirit opened in a way I had never experienced before.

After Mary became available, I spent many weeks working with her—often up to a month a year. She became a close friend as well as a trusted healer. I watched the scars begin to lighten over time. I remember seeing the top letter slowly materialize on the eye chart. When my right eye hemorrhaged in 1981, I was ready for it. I was functionally blind in both eyes for only a matter of weeks. My left eye took over again, and I was able to see! At this time, my left eye is correctable to twenty-twenty (remember, I was told I would *never* regain vision at all), and I am still working on my right eye. The condition is stable, and I have not had a further hemorrhage for years.

What is the definition of self-healing? For me, in this case, it was knowing when to not accept a diagnosis and how to have the perseverance to turn the condition around. It was knowing that I could not do it all alone and that I had to seek out expert help. It was having the faith to persist and to build this faith into my way of life.

UNCOVERING CANCER—
THE CASE OF S.P.

S.P. was a twenty-nine-year-old occupational therapist when she participated in a five-day workshop with fifteen other people. She had tried meditation only once before and thought the whole idea was slightly ridiculous, referring to it as doing "the thing where you sit still and you don't talk."

She describes her feelings during the first theta meditation this way:

I didn't have any expectations, so I was really surprised when I started thinking of my mother and her holding me as a baby. I hadn't been thinking about my mother—I had been thinking about having a child myself.

S.P. knew that her mother had felt abandoned by her father. Because of this she had difficulty connecting with S.P. when she was born three months prematurely. S.P. had little information about the story, except for the memory of seeing photographs of herself in an incubator.

She continued:

The story became very clear from my mother's point of view during the meditation, and I felt all her emotions. Then it switched to my perspective, except that I had arms that were extremely long. I was trying to reach up to hold my mother, but I wasn't able to touch her or anybody else.

I was trying to reach out with something other than my skin. I remember trying to do it with my eyes, but not getting any clarity—everything was foggy. During the meditation I felt an expansion of myself, something that I now call my "energy body," reaching out to the people around me; but I could not use my skin or my eyes or my arms, because everything was muffled in the incubator.

The sadness came when I wanted to reach out to that baby, I wanted to hold myself. That's when Grumbly came up.

Grumbly was a sticky, black, tarlike entity down below her navel. It had hands and feet that were in constant motion, like a Tasmanian devil. It had claws, but it was sticky, so it could take things and devour them, pull them away and consume them.

In the past, whenever S.P. started thinking about family, or love, or having children, she developed anxiety that would escalate until it felt like a panic attack. During the meditation, when she was aware of Grumbly, a voice said, "No, you're not going to be able to have kids. This is something you are not going to be able to do unless you make some changes."

S.P. wasn't sure what those changes were that she needed to make. After the meditation, in the grounding and sharing phase, she talked about the possibility of needing to get more in touch with her own inner child and to learn how to play before she could have her own children.

I remember talking about my desire to have a family. There was shame at having that need, but there was great freedom at being able to talk to a group of people and feel like they were understanding and accepting me.

In the later, self-healing meditation, she asked to speak to the part of herself that knew about the pain and the emotional stuff that was going on inside Grumbly:

I slithered out into the dark thick goo, and I turned into me at about age eight or so. I was just like a kid who had been living in a cave. I was all dirty and scared and dark and quiet, and I crawled out from a crack.

When S.P. saw this imagery, she knew she had to find a way to heal this inner child.

I stuck with that feeling of wanting to have a child. The more I thought about it while I was walking around [the retreat center], the more I realized I needed to get a checkup, because there was going to be something that was going to get in the way of me being happy in that area—having a child and a family. The word "cancer" was always very clear in my mind.

S.P. had gotten a pap smear before the workshop, and the results had been negative.

But my gut feeling was so strong from my stories that were developing from the meditations that within a month, I had several pap smears from different clinics, because I didn't trust the testing in clinics that I was going to.

Sure enough, each doctor said that my cervix looked abnormal, but the test results were coming back normal, so I must be fine.

S.P. still did not trust the results of the pap smears. She continued to practice the meditations she had learned in the workshop at home.

My body—I don't know what to call it—that voice that we get in contact with when we meditate, whatever that part of us is there, the inner knowledge, would not allow me to let it slide.

I went to a female gynecologist and told her the story of my meditations and my intuitions regarding them. She kind of humored me and did a really thorough pap smear.

When the test results came back, both cancer and a precancerous spot were found on S.P.'s cervix.

Three surgeries were required to remove the cancer and return S.P.'s cervix to full health.

> I'm just so glad that I followed my intuition, because it would have gone into my uterus if I had waited, and I would have probably had to have some form of hysterectomy. Through the meditations and discussion that you led, I was able to focus on my body's language and listen well enough to prompt me to seek [one final] pap smear. If I hadn't had another pap, the cancer may have spread to other organs.

Now, nearly a year later, S.P. is using her meditations to get to know who she really is and what she wants out of life—and to heal.

THE HEALING OF OTHERS

This brings us to the other aspect of healing, which, while not as common, is gaining in acceptance in this country. That is the healing of others through the use of laying on of hands, energy balancing, touch, or even by mere presence alone. During my years spent in England, I had the good fortune to get to know many successful people in the healing profession. The research that was carried out with some of these individuals involved monitoring their brainwaves while they healed others. The brainwave patterns of the person being healed was also frequently monitored.

Max Cade conducted much of this London research, examining the brainwaves of many healers in both the British Alliance of Healing Associations and the National Federation of Healers. He also worked extensively with two of Britain's most successful and well-known healers of the time (the 1970s), Rose Gladden and Major Bruce MacManaway. He discusses his findings in *The Awakened Mind*. He found not only that the most notable healers were generally operating with an awakened mind brainwave pattern, but that there is a measurable brainwave change in the patient as well:

After a successful healing session, a patient will be more relaxed yet at the same time more wide awake and better able to respond to emergencies. These physiological changes seem to be the result of the healer inducing in the patient one or both of two responses, namely, deep psychophysiological relaxation and fifth-state consciousness (the Awakened Mind combination of beta, alpha, theta, and delta).[1]

My work in the past fourteen years since my return to the United States has afforded me many opportunities for measuring the brainwaves of people in similar healer/healee relationships. I have looked at the practitioner/client relationships with brainwave monitoring in Feldenkrais work, Tragering, polarity therapy, hypnosis, and so forth, and have seen the practitioner affect the client's brainwaves dramatically.

While I do not teach *healing* per se, I have taught many classes that involve the self-control of energy for a variety of uses. Generally in these classes the students have already undergone brainwave training for the awakened mind state and already have some, if not a lot of, access to preliminary awakened mind states and the understanding of how to manipulate their own brainwaves. The following is a meditation from one of those healing classes.

GATHERING THE HEALING ENERGY FOR TRANSMISSION TO OTHERS

Close your eyes and allow your mind to run over collected memories of
   past meditations. (**)
Remember the depth of meditation that you have experienced.
Begin to take that depth of meditation into you now. (*)
It's almost a sense of moving through different states of consciousness as
   you go down, (**)
down and back, (*)
down and back, (*)
down . . . and . . . back.

(pause for one minute)

As you move deeper into your meditation, you can feel the various
    tensions in your body letting go. (**)
Throughout your body, waves of relaxation and release, (*)
letting go. (***)

Allow any thoughts, images, or outside disturbances to become com-
    pletely irrelevant (*)
as you move into that deeper state which is within you. (**)
Perhaps you find a light there, (*)
or colors, (*)
signposts to help you on your journey, (*)
guidance to take you further, (*)
and deeper, (*)
to a place that opens inside you, (**)
a place of freedom. (*)
This place knows no bounds, (*)
has no limits (*)
other than the limits you choose to impose. (**)
From within this space now, you can begin to contact the healing energy
    that exists in the universe.

(pause for thirty seconds to one minute)

You might sense it as a vibration, (*)
as a light, (*)
as a sound, (*)
or a voice, (**)
as a vision, (*)
or a sensation, (*)
or just a kind of knowing that it is there. (***)
And you can begin to take this healing energy into you, (**)
feeling it running through your body, (**)
through your spirit, (**)
through the very core of your being.

(pause for thirty seconds to one minute)

If there is any particular area within you that needs healing, (*)
concentrate the healing energy into that area. (***)
Become aware! (**)
Notice the changes as the energy moves in you. (**)
Allow the healing energy to move over you and through you, (**)
like a shower of iridescent flashes of light and sparks. (*)
Allow the exquisite sensation to make its way (*)
to your mind, (**)
to your knowledge, (***)
and allow yourself, if you wish, to have a vision.

(pause for one to two minutes)

And again get in touch with your healing energy. (***)
Now, as you draw it into your body, (*)
you may become aware of the source of that energy, or you may not. (*)
It doesn't matter. (*)
But as you draw it in, bring it down through the top of your head, (**)
through your heart, (**)
all the way down to the base of your spine, (**)
to energize and recharge your vital forces. (***)

And as you take in the energy now, (**)
bringing it in through the top of your head and down through your heart,
    (**)
take it also down through your arms, (**)
coursing down your arms. (***)
And you feel your hands becoming energized, (**)
feel your arms as channels, (**)
the energy coming in through the top of your head, down your arms, (*)
moving to your hands, (**)
and out through the palms of your hands. (***)

You might feel it as a tingling. (***)
You might see it in your inner vision radiating outward from your palms.
    (***)
Hold this feeling, (**)
keep it inside you, (*)

so that, in the future, you can open the channels even more easily and effectively.

(pause for up to one minute)

And in your own time (*)
and in your own way, (*)
give your thanks and appreciation for what you have received (***)
and allow yourself to close your meditation (**)
without shutting off to the source of the healing energy.

(pause for thirty seconds to one minute)

When you are ready, allow yourself to return. (***)
Reemerge and reawaken, (*)
arousing yourself and becoming present, back in your room with your eyes open.

This is one among many possible techniques for focusing healing energy. I prefer my students to experience the flow of energy within their own bodies, work with self-healing, and practice manipulating and directing energy within themselves, before they start (if indeed they ever do start) using that energy to positively benefit other people.

The next step might be to place my students in pairs and let them take turns practicing the simple laying on of hands. Then I might begin to encourage them to find the area on their partner's body where their hands are naturally drawn. I ask them to pay particular attention to sensations they are experiencing in their hands. These can range from extreme temperature changes to tingling, itching, throbbing, or pulsating sensations.

When you practice the laying on of hands, you should give your hands free reign. Notice if you are stuck or drawn like a magnet to a particular place. Stay there until the energy is finished—even if you think it is time to move on, if your hand isn't finished, let it remain. Sometimes you may have a gut-level feeling that your work is incomplete; sometimes your hand just doesn't want to move. Follow your intuition.

I later ask my students to be open to, or even look for, images, pictures, or other sensory input that might give them valuable information about their partners' conditions. If it is appropriate, they can dialogue with their partners about what they are experiencing. I ask the individuals receiving energy to try to remain aware throughout and, if it is appropriate, to report on their experiences. If one feels heat coming from his or her partner's hand, it is helpful to the healer in training to have that feedback. It is also useful for the healee to maintain conscious awareness of the changes that are taking place within his or her body, mind, emotions, and energy system. Likewise, sharing any images or sensations he or she may be having could be helpful to the healer.

When I am doing this type of *energy balancing* with clients, I might also ask them to simultaneously develop a dialogue with parts of themselves, as in the second meditation in this chapter. While my hand is on a particularly troublesome spot, I might ask them to go inside and look for the image or other sensory manifestation in the area beneath my hand, identify what the problem is, and then have a conversation with it. This time, when I am asking the questions, I determine my next question from the previous answer. In this way I help them to find out the source of the energetic block in that location and begin to release it.

HEALING CIRCLE

I would like to close this chapter with one last healing meditation. This should follow a deep induction and relaxation and a stilling of the mind.

There is almost a sense of moving backward, (*)
backward into softness, (*)
backward into warmth, (*)
falling, (**)

falling quickly now, faster and faster into a deep trance;
and the faster you fall, the more relaxed you become. (***)

As you fall, you are aware of passing through different levels of consciousness. (**)
They may be presented to you as images, sights, sounds, voices, body
sensations, feelings, even smells and tastes. (**)
And you fall beyond them. (***)

Now you begin to slow. (*)
Still you are falling, but gently, much more slowly, almost drifting down,
(**)
and down below you, far, far down below you, you see a circle; and it is
toward the exact center of this circle that you are falling (**).
A healing circle. (**)
Getting closer and closer now, (*)
and the closer you get, the slower you are falling, (**)
and you see that it is a healing circle, the very center of which you are
moving toward, (**)
now hovering just above it. (***)

And with exquisite gentleness and softness, you come to rest in the exact
center of this circle, (*)
so that you are lying symmetrically in the center of the healing circle,
which you can see or feel very clearly all around you. (*)
And the healing begins. (***)
It may take many forms. (**)
You may feel strange and pleasurable sensations. (**)
And as the healing permeates your body, there may be messages that
come to you, or just an indication of where the next step on your path
may lead you. (***)

And now, ever so gently, you feel yourself rise out of your physical body
into your spirit body, your higher self. (**)
And the healing continues and becomes even stronger as it is directed to
your spirit body, (**)
and it feels good.

(pause for up to one minute)

And now, as your higher self begins to feel whole and full, you descend again, (**)
passing through the emotional and mental bodies, (**)
cleansing them as you move through, (***)
back into the physical body once again, (**)
taking into the physical body all the healing you received in your higher self, (***)
integrating that healing into the physical body, (***)
feeling yourself aligned, integrated, and in harmony within yourself. (***)
At peace.

1. C. Maxwell Cade and Nona Coxhead, *The Awakened Mind* (New York: Delacorte Press/Eleanor Friede, 1979), p. 192.

CREATIVITY, LEARNING, AND THE AWAKENED MIND

Creativity takes many forms. It often involves originating something through your own invention, bringing into being an idea, concept, or object that has never occurred before in quite that way. Creativity is also embellishing, fine-tuning, refining, or altering something already in existence to make it better. Creativity can apply to any area of your life. It might involve finding a new way of cleaning your home, or making a relationship more satisfying, or developing a project at work, or painting a picture. Creativity can manifest in the way you view your world and respond to your environment, how you handle your thoughts and emotions, and even how you develop your spirituality.

There is only one major difference between optimum brainwave states of creativity and deep meditation. In states of creativity, you need to *think*. Brainwave patterns of peak performance are simply brainwaves patterns of meditation with beta, the brainwaves of conscious thought, added to them. The lower-frequency brainwaves remain very much the same. A pattern of creativity is merely one of the many forms of awakened mind patterns. Examples can be seen in Figure 1. The patterns vary, depending on both the activity and the signature pattern of the individual.

Producing the right amount of beta for the creative state you want is not quite so simple as it seems. The beta needs to be added in the right frequencies and the right quantity for the purpose that you

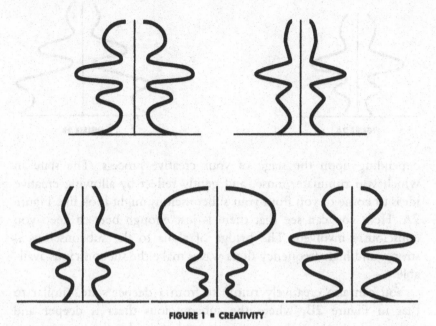

FIGURE 1 ▪ CREATIVITY

intend. High-amplitude, splayed beta does not give you the optimum creative states that rounded beta peaking in the midrange gives you. For this reason, when we are training for states of creativity, it is best to do so in three steps:

—First it is advisable to learn to reduce *all* your beta in order to clear out those higher frequencies that are so common in a stressed-out mind.
—The next step is to practice meditations that teach you to access the important ingredients of alpha and theta.
—Finally, you can add the beta back to your meditation pattern to create the conscious content that you need for your creative endeavors.

As we have seen so far in this book, this is the master plan for developing any awakened mind pattern. The awakened mind pattern of creativity might look similar or identical to an awakened mind pattern being used for self-healing. Only the content will be different. The amount of beta you are producing might also increase,

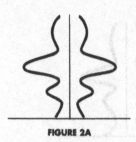

FIGURE 2A

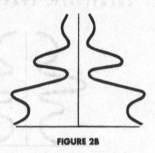

FIGURE 2B

depending upon the stage of your creative process. The state in which you ruminate, muse, and gently reflect by allowing creative ideas to come to you from your subconscious might look like Figure 2A. Here you can see that there is just enough beta to keep you consciously involved. The bridge of alpha to the subconscious is strong, and high-frequency theta waves make the subconscious available.

You can also creatively ruminate from a deeper state, similar to that in Figure 2B, where the subconscious theta is deeper and stronger. Here the alpha is strong enough to bridge the gap between the subconscious and conscious minds. The beta is active enough to permit you only conscious awareness, but no conscious direction of your pondering. This is the type of state you experience in the morning when you have just woken up and still have access to the contents of your subconscious mind.

Light musing often begins from a conscious state. From beta, you add alpha, theta, and delta and move *down* into a state of creative reflection. Figure 2B shows the opposite. You might start from a state of meditation and then move *up*, adding beta waves to awaken your mind and to muse creatively, or deeper musing may begin from an unconscious state (i.e., sleep). You may find yourself slowly waking up, contemplating a particular issue, as if you had been mulling it over in your sleep. Very gently and slowly consciousness creeps in. You may feel as if you gradually begin to overhear a conversation already in progress. If you can stay in that state of deep musing, you can have a long and meaningful interaction with your subconscious.

EARLY-MORNING STATES
OF CREATIVITY

You don't have to decide the subject the night before to make use of the early-hour creative time—especially if it is strong within your subconscious. Processing takes place during your sleep. First thing in the morning, your brainwaves are wide open—theta, then alpha, then beta. You can simply scan what is deep inside and bring it up to your conscious mind.

CHRISTMAS CARDS

An old friend of mine called Jennifer told me the story of her Christmas cards. For weeks she had been putting off even thinking about writing "the blasted things." Deep down inside she knew that there were some old and dear friends she wanted to communicate with and that this was an excellent opportunity to rebuild a number of broken bridges. But she was overwhelmed and confused as to where to even begin. So she avoided taking any action until the whole issue became a thorn in her side and began to put a pall on her holiday. She halfheartedly looked at cards in a number of stores, never finding any she liked well enough to buy.

Jennifer told me that in a semiwaking state in the early morning, when her mind was completely relaxed and open, she saw herself writing the cards. In that instant she knew exactly whom to write to and what to say. Next she remembered where she had stored her unused cards from last year. (Jennifer told me this was a long-term affliction—she had bought, but not written, cards for several years).

That day she effortlessly wrote cards to more than a dozen friends from the past, putting her heart into it and enjoying the feeling of reconnection. When I telephoned to thank her for my card, she told me of her early-morning reverie and exclaimed, "I feel like my Christmas card block is gone forever."

HOW TO USE EARLY-MORNING REVERIE

If you take advantage of your own early-morning reveries, you might find yourself retrieving ideas from your subconscious that had previously eluded you, or coming up with inspirations and new ideas that you did not expect. If you set a goal for yourself, or have a particular project you are working on in a very focused manner, you can actually program your mind to use your early-morning state for maximum creativity. To take the fullest advantage of this state, try not to increase the low amplitude of the beta you awaken with. Simply place the object of your desired creativity firmly in your mind and then *do not think*. In other words, use your beta waves only to the extent that you give yourself a conscious context for your reflections.

If you introduce the subject that is important to you, or your mental block about how to proceed with something, you can receive answers and inspiration immediately. Just remember, you need to catch this state before your channels close up.

The next vital element to this form of creative exploration is *time*. You cannot reflect in this way easily under the pressure of an impending alarm clock. You need time to be able to let the burgeoning contents of your theta waves gently bubble to the surface. When beginning this practice, choose a day when you have no need for a specific rising time, such as a weekend. Once you get used to using these early-morning states, you can actually tell yourself to wake up a half hour before your alarm goes off in order to be able to use this precious time on a daily basis.

The final important component to early-morning creativity is to find a way to retain the ideas and information that come to you without waking up or overactivating your beta waves, which would immediately reduce your theta waves. When that happens, the process is over. As in some of the previous meditations, the trick is to find a key—a symbol, word, concept, image—something that represents to you the information that you are receiving from your theta, *without going into detail*. Just as you did when you awakened from a meditation, bring this key into your beta when you awaken from your morning reverie. In this manner, you will have access to the memory of your experience when you wake up fully.

It might take practice to learn to make the best use of those deep states of inner reflection, but it is well worth the effort—or lack of effort, in this case. Effort would only sabotage your goals of *allowing* this experience to happen. These morning reveries cannot be forced. The more mornings you awaken with the *intention* of accessing deep and even hidden creativity, the more natural this process will become. If, on the other hand, you desire these states but your life is ruled by the clock, and you never let yourself stay in them for any length of time because there is always something more important to do, then you are missing the opportunity of communing with yourself creatively and the benefits you might acquire from it.

One final, important note here. When you receive a creative inspiration in this way from the depths of your subconscious, *act on it*. This will reinforce the positive benefits of being in this state, which will in turn encourage your psyche to keep re-creating these benefits. If you have constant inspiration that goes unheeded, you are creating negative feedback for yourself, in essence telling yourself that the information that you receive from these states is unimportant and not worth acting upon. This nonaction could result in a desensitization process that might eventually hinder you from having such inspirations. Reinforcing the benefit of the creative messages that come to you from your theta waves will help them to continue to flow fluidly and more frequently.

I have used this early-morning creative time for working on this book. Often without even intending to, I have gone to bed with a certain chapter or section on my mind. Several times I have awakened in the middle of the night or in the very early morning hours to "hear" that section of the book being written. I grab a pencil and take notes.

I have learned I have to write fast and stay in a trance state to retain the detail of all of the information gushing out of my subconscious. If I sit up and turn on the light so that I can write more legibly, I take a chance of losing my memory of what I was hearing. After missing several of these insights, or at least not retaining the exact series of words that I wanted, I began to place a paper and pencil by my bed every night just in case. I do that to this day, never knowing what might occur to me in the depths of my nighttime reverie. My only difficulty is occasionally deciphering the scrawl I find by my bed in the morning.

Occasionally I might awaken with the vague memory that something occurred to me while I was half asleep, but I won't remember what it was until I look at the notes I took during the night. Then, usually, the whole reverie returns as if by magic. If you are new at this, I suggest that you take time to sit down with your notes and fill in the details as soon as possible after awakening. The memory of the experience returns with more clarity the earlier you flesh it out.

CONSCIOUS CREATIVITY

The type of creative state that I have described above is a very passive state. While practicing this state, your main conscious task is to *allow* information, ideas, and creativity to rise from within you, and to find a way to lodge them, once they have arrived, within your conscious grasp. Other forms of the creative state are more active. The more active the state is *consciously*, the more beta is involved. If you want to intentionally direct your creative flow, add more beta to your awakened mind pattern, see Figure 3. (This portrays more and more beta being added to the creative pattern.) The working meditations that follow, similar to the self-healing meditations found in Chapter Five but with the focus on creativity and problem solving, help develop what Max Cade called "mental fluency," the ability to manipulate the contents of the mind. These contents come in many forms: verbal, emotional, musical, mathematical, sensual (visual, auditory, kinesthetic, tactile, olfactory, and gustatory). Developing mental fluency creates "flexibility of imagination for stimulating creativity, for loosening one from the tight grip of the so-called normal state which we are conditioned to believe is the most desirable, imposing its rigid limits on our self-development."[1]

Practicing altering the contents of your mind not only helps develop optimum brainwave states, as we have seen in earlier chapters, but also helps your creative capacities and potentials to unfold and flourish. State helps content and content helps state at the same time.

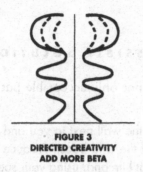

FIGURE 3
DIRECTED CREATIVITY
ADD MORE BETA

We will begin our practice of mental fluency and problem-solving meditations with no specific goal other than to get the creative juices flowing. Later we will focus on solving specific problems.

The goal of the following meditation is not to help you solve a real-life problem but to teach you to allow yourself to be as wildly creative as possible. Do not allow your ideas or solutions to be limited by normal reality. Think of two or three imaginative solutions to each situation. You may wish to prepare for this exercise by spending some time relaxing, stilling your mind, and entering a state of meditation.

MENTAL FLUENCY MEDITATION

Find creative solutions to the following situations:

1. You are in the middle of a long, narrow alley with walls on both sides. The entrance to the alley is far behind you; the exit is way in the distance. Suddenly, you are facing a menacing wild lion that has come out of nowhere . . .

2. You are standing on the edge of a steep gorge, looking down over the precipice at the jagged rocks far down below you. To the right and left are impassable boulders blocking your way. Behind you is a sheer, vertical rise . . . .

3. You find yourself lost in a foreign country. You do not speak the language and you have no money. It is getting dark and you are hungry . . .

POSSIBLE SOLUTIONS

Using your fantasy is not only acceptable but advisable!

1. —Create a door in the wall next to you and quickly slip away.
   —Become master of the wild beast. Produce food and tame him.
   —Point your finger at him and, using your special powers, shrink him
   to the size of a pussycat.
   —See God in the wild lion and contact him on a spiritual level, asking
   him to be your ally and protector.

2. —Cause the gorge to fill with water, and swim to the other side.
   —Levitate to the top of the sheer, vertical rise, and walk away.
   —Jump off, but allow yourself to float gently down to the bottom,
   landing with featherlike lightness.
   —Using your mind power, create a bridge and walk across.

3. —Know exactly where to look for a lost $100 bill.
   —Round the corner and run into an old friend, who just happens to
   have moved into town.
   —Find a church, synagogue, temple, or holy place of your choosing,
   enter, and turn your problems over to your higher power, knowing all
   will be well.
   —Decide to be a street performer and sing for your money. Someone
   immediately gives you a small fortune.

Your solutions can range from the completely fantastic, like most
of these, to the more potentially realistic. The more possible alternative resolutions you are able to find, the more you will be flexing the
muscles of your creativity and developing mental fluency. You can
make up other problems and practice finding solutions, gradually
beginning to substitute real problems in your life for imaginary ones,
and realistic solutions for totally impossible ones. You may be surprised by what you find is possible.

PEAK PERFORMANCE AND PEAK
EXPERIENCE
"Ah-ha! I got it! But what have I *got*?"

Though some people have never heard of the term " 'ah-ha' experience," almost everyone at some time during his or her life has undergone this experience of illumination. While you are in this state, it is unmistakable, but unfortunately, it is often all too quickly gone, as if it had never occurred. The next time it happens we think, "Oh yes! I remember this!" only to forget the experience all over again.

Though an aspect of what is commonly known as *peak experience,* the "ah-ha" phenomenon differs in that it is briefer, a momentary flash of intense awareness. It is the instant understanding that might occur in the midst of a laborious research project, or the moment of exhilaration and unity that might come to you during a beautiful, perfect day on the ski slopes. Peak experience is the "ah-ha" stretched for longer and longer periods, possibly for minutes or even hours.

When you feel an "ah-ha" experience hit you, it is important not to let it go unnoticed or unheeded. When you identify these types of experiences, it becomes more likely that you will have another one. Sometimes the recognition and acknowledgment of an "ah-ha" experience will cause another one to happen immediately, which may then cause even another, and another. I call this the *"ah-ha" loop.*

I have often used biofeedback to successfully identify and perpetuate this kind of occurrence. This method of training the awakened mind is done consciously and with open eyes. There are no deep relaxations and no breakdown of the individual components of the brainwave pattern. First I hook a client up to the Mind Mirror; then I begin involving him in a discussion about something he finds interesting and stimulating. The "ah-ha" experience can then be encouraged in one of two ways.

In the first, the experience comes directly from the client. I may gently prod him by getting him to talk to me about his life's purpose or goal, his spirituality, and his concept of higher power, or his latest creative project or undertaking. Hopefully, from here he will reach a creative peak that I can point out to him. One of the aspects of the brief burst of beta, alpha, theta, and delta that characterize a developing awakened mind is that frequently the pattern goes unrecognized and unnoticed. Once it is pointed out, *the individual recognizes it for what it is*. It is this recognition that causes the recurrence of the pattern. The longer this loop can be sustained, the more experienced the individual will become in recognizing and identifying the state, and the more easily it can be reproduced in the future.

Many times I have seen clients have bursts of awakened mind patterns and not realize the validity of what they were experiencing. When I stop someone during a session when this occurs and say, "What was that?" the answer might be, "Oh, I just had an idea, but it was nothing." I then know to pursue the idea that was "nothing," because I know that it came from a creative state of mind.

The following is an example from a therapeutic session with a client. R.N. was interested in discovering the childhood root of an emotional problem he had been carrying for many years. He insisted that it originated with his mother and there was nothing wrong between his father and him. He was showing a typical splayed beta brainwave pattern with little alpha and no theta. Quite unexpectedly, during the conversation about his mother, R.N. produced a beautiful, full awakened mind pattern for less than a second. I stopped him instantly and asked him what just happened. He looked so taken aback that I knew something had just happened, and he replied, "I just had a picture of my father beating me." Obviously, we turned the topic of exploration to his father.

The memory occurred to him as an instantaneous flash and was gone. If I had not pointed it out to him, it is likely that he would have ignored the picture while burying it even deeper in his subconscious. Identifying this "ah-ha" for R.N. opened up a whole new avenue of development for him.

In the same way, people often ignore their flashes of brilliance, assuming that they just are not capable of creativity, so the flash of insight they just had must not be relevant, or there must be a hole in it somewhere, or they aren't good enough to carry it out, or . . . An insight can come bursting into consciousness, and because an individual has not been encouraged to believe in her flashes of inspiration, she may not recognize it for what it is; or even if she recognizes it, she may not act upon it.

One client, G.D., was beleaguered by creating an Easter pageant for her church. She had overcommitted and underdelegated, and the result was high anxiety, a feeling of being overwhelmed, and illness. She came to me not knowing what to do and feeling that she was up against the impossible. I let her talk for some time and then asked her to think about solutions. In the midst of her complaints and insistence that there were no solutions, she had a strong awakened mind flare. I stopped her midsentence and asked her, "What did you just think?" She said, "Oh, but I could never do that!" I asked her what it was that she could never do. At first she was very resistant to even discussing it, but when I pointed out her brainwave pattern to her, she relented. G.D. explained that a major part of the stage setting that she had to make by hand might not really be as necessary as she thought, and that perhaps she could do without it, thereby cutting her workload in half. It took the rest of our session for G.D. to revise her to-do list so that she could reduce her labor while keeping her commitments and feeling satisfied with the resolutions. The following week G.D. came in relaxed, satisfied, and proud to report a very successful Easter pageant.

What if I can't get an individual to produce a brief awakened mind pattern by talking about his goals, purpose, ideals, or inspirations? A second way I can bring about an "ah-ha" experience is by pointing out something about the individual's state of consciousness and describing it in the language of brainwaves. This will often spark a moment of illumination, which can then be developed into the "ah-ha" loop.

So how does this stimulate an awakened mind? Everyone *knows his or her own consciousness better than anyone else.* Usually people have never tried to use words to describe how their minds are working. Yet they might be in my office in order to get help with that very thing. I teach them a new language. I give them a vocabulary for describing their states of consciousness. Then I look at their own individual state of consciousness on the EEG and describe it to them, using the vocabulary I have taught them but personalizing it to their own experience.

Hearing their own state of consciousness fed back to them in this way can be mind-boggling. Hearing about your own state of consciousness in real time, while it is happening right in front of you, creates an "ah-ha" experience. I use that "ah-ha" to stimulate another "ah-ha," and on and on. Often people can sustain as many as a dozen awakened mind flares in this way until the mind tires of it and the normal state of consciousness reasserts itself and begins to digest the experience.

So what has just happened? When the brainwaves flare into a brief awakened mind in this way, the conscious, subconscious, and unconscious mind all become open to one another without constraints and bottlenecks. The flow of information is able to move in all directions at once. The learning experience embodied by the beta waves is transferred instantaneously to the subconscious, where it recognizes the truth of the learning on a deeper level. This recognition of deep inner truth is instantly transferred to the conscious mind, and the conscious, subconscious, and unconscious meet and fuse in a moment of all inclusive understanding and "ah-ha!"

I was in the office of the CEO of a major British corporation, taking him through a brainwave profile. We began talking about how his spiritual process related to his work ethics, and he produced a strong awakened mind pattern. I froze the pattern on the screen, showed it to him, and then released it. He said, "You mean, that's it?!" When I said yes, he immediately produced another one. This happened twelve times before he settled down. Then we began to work on how he could produce the same pattern intentionally.

There is a slight difference, however, between the brainwaves of the "ah-ha" and the brainwaves of an ordinary awakened mind. The "ah-ha" experience tends to come in brief bursts or flares of *high amplitude and very strong* awakened mind patterns. If you were to maintain this pattern into a continuous awakened mind, it would follow that you would be continuously living in a peak experience. And in a way you would. Figure 4A shows what the pattern might look like immediately preceding the "ah-ha." Figure 4B shows a burst lasting anywhere from one second to one minute before returning to approximate the previous pattern. What characterizes this awakened mind pattern is the extreme amplitude of the alpha and especially the *theta* brainwaves. This high amplitude is rarely main-

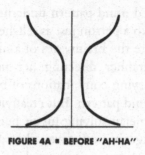

FIGURE 4A ▪ BEFORE "AH-HA"

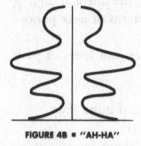

FIGURE 4B ▪ "AH-HA"

FIGURE 4C ▪ PEAK PERFORMANCE

tained over time. It is something in the burst itself that gives the individual the experience of the onset of illumination.

PEAK PERFORMANCE

As I mentioned above, peak performance differs from the "ah-ha" experience in duration, the way it develops, and the intensity of certain brainwave frequencies. The onset is not an instantaneous burst, but rather a developed stable pattern; the alpha and theta are of lower amplitude; the theta amplitude is lower in relationship to the alpha amplitude. In other words, the subconscious is not as open as in the "ah-ha," and the channel for the flow of information is not as strong as in that brief period of extreme illumination (see Figure 4C). There is one more variable: beta fluctuates in both frequency and amplitude during peak performance states in accordance with the activities being performed.

Because the awakened mind pattern underlies peak performance, this state can be taught to a person just as self-healing can. I once had opportunities to measure the brainwaves of a musician while he was composing, a choreographer directing her new dance, a dancer dancing. All showed varying combinations of beta, alpha, theta, and delta in an awakened mind pattern. But creativity does not stop with the arts. The mathematician solving his or her most difficult problems, the accountant figuring out taxes, the scientist working on a new theory, the athlete competing or working for his personal best, the skier having fun on vacation, the homemaker cleaning and decorating, the executive attending a board meeting, the florist arranging flowers all may be producing the brainwave patterns of peak performance.

So how can we improve our abilities to enter into states of peak performance?

1. Train your brainwaves to develop an awakened mind state.
2. Practice meditation regularly.
3. Learn focus and concentration while undertaking an activity.
4. Set an intention before beginning the activity.
5. Practice awakened mind techniques, exercises, and meditations immediately before undertaking an activity.
6. Give yourself positive feedback every time you experience the state of peak performance, no matter how weakly or for how brief a time.

Don't necessarily expect your peak performance to occur initially in the area where you want it most. Be open to the experience of peak performance *anywhere* in your life. Of course you want to excel in your chosen field, but that will come. Anytime you *are* experiencing peak performance in any area, even a hobby, sport, or activity with little importance to you, give that experience the validation, respect, and positive feedback that you would give an experience in your chosen field. Detach the *state* from the *content*.

INTERNAL FACTORS OF
PEAK PERFORMANCE

Without considering the content, try to notice what the state feels like. Be aware of how your body feels, of what is happening with your emotions, of how your mind is operating. Notice how you are breathing, what your focus is, and how it feels; how you are sitting, standing, or moving.

Find ways of grounding the experience just as you do after a meditation. Find symbols, images, or words that describe the felt sense you have inside. Crystallize these experiences into a few keys for yourself. Then try to recreate the experience by calling forth those keys. Reenter that state of consciousness as if you are finding a switch inside. If you find the right switch, you can flip it on. For some people this experience is more like turning a rheostat down or up. Whatever image you choose, the essence will be in finding your own internal key and then using it to unlock the door to your personal higher states of consciousness.

EXTERNAL FACTORS OF PEAK
PERFORMANCE

Notice what day of the week it is, what time of day it is, where you are, what the temperature is, how long it's been since you ate, how much sleep you had last night, and any particular occurrence that seemed to help spark your episode of peak performance. Be your own researcher. Every time you have an experience of peak performance, notice the variables. After a few times of doing this, you will become aware of the similarities and differences. This is invaluable information to help you set up the right conditions to make your periods of peak performance more frequent, longer, and more focused on the content that you wish.

For example, if your periods of peak performance always come on an empty stomach, what does that tell you about having a big lunch right before an important afternoon of creativity? If your peak performance is always at a particular time of day, then that is an indication of when you might want to plan to do your most important work. Using the internal keys and the external factors in combination can begin to help you create your own peak performance times whenever you wish.

Meditation on a particular project or activity can help you stimulate the appropriate brainwave states for peak performance as well as stimulate the appropriate content. You may wish to practice the following meditation after a period of relaxation and mind clearing. Remember that this is a *working meditation,* so you don't want to get relaxed to the point of losing conscious awareness. This meditation can take any length of time, from a few minutes to an hour, depending on how much detail you wish to experience and how long you wish to stay in each phase of it.

The format of this meditation is generic—I have used multiple choice where appropriate. Please adapt it to suit your particular needs. The more specific and detailed you can become in both your imagery and other content, the more effective the meditation will be.

PERSONAL CREATIVITY MEDITATION

—Begin by allowing your mind to clear of all thoughts.

—Put aside any thoughts of creativity, the project that you are working on, or any block you may have encountered.

—Allow yourself to become very present in this moment, and simply focus on your breathing.

—Remember what it feels like to be relaxed . . . what it feels like to be deeply at peace inside yourself . . . and withdraw yourself into that place.

—Remember what it feels like to let go of your beta brainwaves . . . and to let go of any stress in your body.

—Scan your body to make sure that you are not holding any tension.

—Make sure that your head and facial muscles are relaxed . . . your

neck and shoulders . . . arms and hands . . . chest and stomach
. . . back and spine . . . hips and pelvis . . . legs and feet . . .
—If you have any difficulty relaxing certain areas, simply breathe into
those places and release any tension as you exhale.

—Now experience your *whole body* as relaxed . . .
—your mind as quiet and still . . .
—your emotions as calm and clear . . .
—your spirit as peaceful . . .

—From this place of inner serenity, harmony, and relaxation, you can
begin to allow some images to occur in your mind.
—Imagine or create an environment appropriate for what you wish to
create. It could be an office or study, a workroom or shop, a studio or
gallery, a kitchen or salon, a stadium or playing field, a classroom or
auditorium, theater or hall, swimming pool or ski slope . . .
—Design the best environment possible for developing your creativity.
—Make this environment safe, comfortable, attractive, relaxing, and
stimulating.
—Sensualize the environment . . .
—Walk around it . . . notice the colors, shapes, and forms . . . the
textures and structures . . . the sounds and smells and even tastes.
—If it is indoors, what kind of furnishings and equipment does it have?
. . . what kind of lighting? . . . does it have windows? If so, what is
the view, and can the windows be shaded if you wish?
—If it is outdoors, what is the scenery? . . . the atmosphere? . . . temperature and weather? . . . time of day?
—Are there other people involved in what you are creating? If so, what is
their part? Where are they placed in this space? How do they relate to
you?

—Whatever space you wish to create in, allow it to be the best possible
space for your purposes.

—Now place yourself in this space . . .
—For your optimum creativity, how are you dressed? Do you have any
special clothing (a smock or apron, a uniform, a costume, formal ap-

parel, or just comfortable clothes)? Imagine yourself dressed exactly the right way for your purposes.

—Walk around your environment. Enjoy it. Make any changes that you want to make. Using all of your senses, get yourself even closer to this creative space of yours.

In this part of the meditation you have been using sensualization to help you access and develop alpha waves, which are an essential component of the brainwave pattern of highest creativity. At the same time, you may have found out some essential information about your working environment and what your needs are. It is now time to take the meditation deeper and allow yourself to develop the theta brainwaves of the creative storehouse. For that purpose we now enter a state deep inside.

—Allow your space to become very still. If there are other people moving about, bring the scene slowly to a freeze-frame snapshot of sensory imagery in your mind.
—Find a comfortable place to sit down in your space, and allow yourself to meditate.

—In this meditation, feel yourself going very deeply inside . . . almost as if you are going down a deep well . . .
—Down . . . and in . . .

—Perhaps you can remember a time of peak experience in the past, a time when your creativity flowed and you *knew* what it was that you had to do . . . and you knew how to do it.
—Remember what that felt like in your mind . . .
—Remember what that felt like in your body . . .
—Remember what that felt like in your emotions . . .
—Remember what that felt like in your spirit . . . the excitement and inner joy of the creation . . . the pleasure in accomplishment . . . the sensation of insight.
—If you have this felt sense stored in your memory, draw it out now.
—Let yourself reenter that state.

—And still you are going deep inside . . .
—deeper and deeper . . .

—There is an inner place where all things are known and understood . . .
—a place of clarity . . .
—a place of light . . .
—a place of understanding . . .
—a place of wisdom . . .
—a place of lucidity . . .
—a place of ease . . .

—And you begin to settle into this place and feel a sense of opening
within yourself . . .
—And you become receptive to what it is that you are wanting to create.
You become open to the possibilities . . . to the simplicity . . . to
the elegance . . . to the profundity . . .
—You begin, on a deep level, to comprehend . . .
. . . what it is that you need to do,
. . . what it is that you need to say,
. . . what it is that you need to feel,
. . . what it is that you need to know,
. . . what it is that you need to experience,
. . . what it is that you need to express,
in order to create what it is that you want to create.

—And you begin to allow yourself to visualize . . . to sensualize . . .
to imagine . . . and to create, in your mind, what it is that you want
to create.

—You imagine it happening . . .
—You experience it as if it were real . . . indeed, it is real as you let
your creativity take over.

—The plan unfolds in front of you.
—There is no effort, only ease, as you see yourself accomplishing what
you want to accomplish.
—It all becomes clear . . .

—Perhaps there was a missing piece. That piece comes into place now.

—Perhaps there was a place of blockage, a place where you were stuck in your creation. The solution or solutions appear . . . as if out of the blue.

—You can feel the movement occurring within you . . . and it feels good . . .

—You may want to spend some time now just being with your creative project, taking it further, letting it unfold, letting it speak to you, listening to it and developing it, whatever it is that you know you need to develop . . .

Let yourself create . . .

(Take as much time as you want here to allow the creativity to come through and to develop.)

—And soon it is going to be time to return to the outside world to put your meditation into practice.

—Very slowly begin to find a closure for your meditation.

—It is very important that you *remember* your experiences in this meditation and that you bring them back with you.

—First take a few moments to complete any unfinished business that you are working on . . .

—Then let yourself crystalize and embody any discoveries or new ideas that you had.

—Find images that represent the creative possibilities that you were exploring, and fix these images in your mind.

—Notice what your body and mind feel like, so that the next time you want to reenter this state, you will simply need to recall this feeling.

—When you are ready, once again become aware of the creative environment that you built around you.

—Allow yourself to return to that place that you imagined so clearly . . .

—And bring your memories from the meditation back with you.

—Take a few moments to contemplate your creative experience in the meditation.

—Are there any changes you need or want to make for yourself to allow the fullness and reality of this inner experience and vision to manifest for you in the outside world?

—If so, imagine yourself making these changes . . . easily, clearly, and consciously.

—What needs to happen in your outside life to allow this creativity to fully materialize?

—Imagine that happening using as many senses as possible.

—When you are ready, allow yourself to verbalize and lock into your beta waves the experiences that you had of your creativity.

—While they are still fresh in your mind, write them down, speak them into a tape recorder, draw them, or tell them to another person.

—*Make your creativity real in the outside world.*

—Know that you can always return to this space inside, to rest, to relax, to become clear and centered, and to create . . .

As soon as possible, implement your creativity. Take the time and the effort to manifest the results of this meditation.

WHAT TO DO IF YOU GET STUCK . . .

Most of us will occasionally experience a time when nothing seems to move us beyond our impasse. We are at a standstill, perhaps baffled or bewildered by being so stymied, perhaps bored or angry, or simply distracted.

You can treat your creative deadlock as any other issue to be healed or transformed.

Find an image for it and dialogue with the image.
Find the positive purpose for the obstacle and the necessary steps to take to clear the blockage.

You can also address the block by sensualizing the issue surrounding it. Approach it from all angles, openly, without expectation or judgment. Walk around it in your mind. Explore it fluidly. Flow with it, under it, over it, through it. Experience it in its different aspects or perspectives until you find a tiny opening, a place where you can work your way inside to make changes.

Let your mind roam freely with these images and see what happens. Energies may mobilize that you never expected you had. New perspectives can open up. Decisions can be made that change the course of events dramatically and unexpectedly. Develop mental fluency, the ability to manipulate the contents of your mind.

Following are some examples from my work that illustrate how accessing creativity can work effectively.

THE REAL ESTATE EXAM

One of my long-term clients failed his real estate exam two times in a row. He was excellent at networking and at sales, but fell short on the mathematical understanding needed to complete sales. M.C. knew the material and could figure out the necessary numbers when he was with a client, but he froze on the exams and could not do the calculations within the allotted time. In our preparatory sessions before his third, and final, opportunity, we determined that the stress of failure caused him to be so tense that he was unable to think properly.

We needed a key that would keep him relaxed. In a deep meditation, he asked his subconscious for just such a key. The image that he saw was a professional-looking briefcase, a rolled-up newspaper, and an umbrella. He practiced accessing that image when he was studying for the exam to help stay relaxed. On the day of the exam, M.C. felt very nervous, uncertain, and insecure. He later reported thinking that it wasn't going to work. But when the calculation part came and his nervousness escalated, he stopped, closed his eyes, and imagined his now familiar briefcase, newspaper, and umbrella. With visions of all the success he would have in his profession as a real estate agent, he completed the mathematical portion of the test and went on to pass with flying colors. And a simple image helped him do it.

PLANNING A PRESENTATION

Dr. T.H. was invited to introduce his research at a prestigious scientific conference being held in Hawaii. As this was his first exposure to a sought-after network of scientists, he was understandably nervous about his upcoming presentation. While secure in his material, Dr. H. did not know how much of it to present and in what detail. He was also uncertain about the style of demonstration and how to use visual aids. His anxiety about the need for perfection was getting in the way of his ability to solve these basic problems. With the time of the conference drawing closer and closer, he called me in distress.

I instructed him to forget the conference, go into a deep meditation, clear his mind completely, and enter a state of peace. Next, I asked him to sensualize his presentation at the conference from start to finish. Because he was so deeply relaxed and felt no pressure or performance anxiety, Dr. H. was able to imagine very clearly just what he wanted to say and the style in which he could best explain his innovative ideas. On completing his meditation, I suggested, he should immediately write down an outline of his presentation and prepare it that way.

When he returned from the conference, Dr. H. was proud to report to me that not only did his demonstration run smoothly and easily, but his acceptance by the scientific community about which he had been so apprehensive exceeded his wildest dreams.

INSTANT ACCESS—A PERSONAL EXPERIENCE

There are times when you might need to know a creative solution to a problem *immediately*. You don't have time to sleep on it or even to relax and meditate for a few minutes. You have to access a creative solution instantly.

I have found this especially true in my more therapeutic work. I occasionally work with clients who are dealing with devastating, traumatic life experiences. In some of these circumstances, the exact word or phrase that I say at a given time could be vital to the success of their process.

On one occasion I was working in Germany with a student who was in deep regression, reliving the Holocaust. While timing in this kind of work is often of the essence, normally I would know what to say and when to say it. The catch in this situation was that he did not speak English and I did not speak German. I was working with a translator who repeated everything I said in German and then told me everything my student said in English—a laborious process at best. Usually this worked well because the translator was highly skilled and used to my work.

In this particular situation, however, my student was experiencing awakening from a very painful deep trance. I knew that the next words he heard had to be exactly right. I did not trust the preciseness of my interpreter's translation, nor was I even sure what words needed to be said. I was also apprehensive of the disturbance hearing my English might cause him. Without thinking, I took a slow, deep breath, awakened my deepest subconscious, and heard myself speak to him in German. He opened his eyes, smiled broadly, shed a few tears of joy, and hugged me. To this day, I do not know exactly what I said; I just know it was the right thing to say.

CREATIVITY IN THE SHOWER

Many people have reported to me that one of their favorite times to meditate is during their morning shower. I relate well to this, since I have been creatively using that time of guaranteed solitude and peace for years. Standing under the warm rain of healing water, it is easy to clear my mind of all thoughts. I take the opportunity to connect with my own spiritual source and offer my day for the benefit of whatever higher purpose is appropriate. Then I allow a vision of how the day can best unfold to enter my mind. I might see pictures, hear conversations, experience body sensations, or use other senses to create an image of the best path before me.

If I am facing a particularly challenging situation, I will use this time to bring a sense of light and peace to the circumstances and ask for guidance in the best way to proceed. The shower becomes a time of healing and regeneration as well as creativity and planning.

CREATIVITY REQUIRING OPEN EYES— THE ARTIST IN US

People whose creative activities actually involve using their eyesight as an integrated part of their creative process can often experience alpha and even theta brainwaves with their eyes open. I have found this among the visual artists I have measured. They often have a greater ability to produce open-eyed alpha while engaged in an artistic activity, like painting.

To enhance this creativity, the artist has two choices: meditate *before* creating, as in the above meditation, or encourage the awakened mind brainwave pattern with the eyes open while in the act of creating.

OPEN-EYED ALPHA

If you wish to create with your eyes open, begin by stopping for a few moments and closing your eyes in preparation. Focus inward and center yourself. Just as in the concentration exercises in Chapter Four, beginning on page 89, send your awareness inside. Let your mind clear of thoughts; focus on your breathing; and become still.

When you are ready, very slowly open your eyes just slightly, to allow light into your field of vision. Feel yourself *bringing in* awareness from the outside world rather than *going out* to meet it. Now, allow your eyes to open slightly more to bring in shape, form, and color, still without consciously identifying or labeling the images that are forming.

Next, allow your gaze to settle on the object of your creativity. *Allow the image to come to you.* Do not go out to get it; rather, bring it inside you, maintaining awareness of yourself from the inside.

Allow yourself to go deeper inside—to the hidden recesses and well-springs of your creativity. This will enable you to begin to access your *theta* with your eyes open. Reach into your depths while bringing in the object of your creativity . . .

And now reach out and begin to create. Feel the connection between your inner depth and your outer expression. Keep your mind, your conscious thoughts, and your higher-frequency beta out of the way. Don't analyze; don't judge; don't expect. Simply *be with* your creativity. Let the interaction between what's inside and how it is manifesting outside flow freely. Get a sense of that movement, and keep it going. And above all, enjoy the state that you are in and the experience of creating.

If at any time you feel you are losing your flow, stop and refocus inside—even closing your eyes if it helps. Then once again, merge with the object of your creativity, and let the movement accelerate.

You can try this with a painting or sculpture, a design or plan, a computer screen or writing pad, or anything else that needs your visual attention while you are working.

BRAINWAVE TRAINING IN THE OFFICE

Meditation in the office is becoming a more accepted practice. Yet I know many people who do not feel comfortable meditating at their desks or in the office lounge. Often they will take a walk at lunchtime and meditate in the park, or even go sit in their car for a few minutes. One of my students put the seat back in her car and went out at noon *every day* for a half hour to lie back and meditate.

While this is a helpful, healthy, and ultimately creative practice, I look forward to the time when companies will provide meditation rooms. If they don't wish to call them such, they could be labeled quiet spaces, stress management rooms, resting centers, or even creativity corners. The lighting should be low, with gentle music playing in the background and an understanding that no one should talk.

A variety of styles of comfortable furniture, ranging from reclining chairs to meditation cushions on the floor or on raised daises, would complete the picture. Fresh flowers and well-chosen artwork would be an added bonus. Contrary to feeling embarrassment every time they close their eyes and meditate, office workers could be encouraged to take some time in the meditation room to consider a particular problem or to regenerate after a difficult meeting. Ideally, brainwave training and biofeedback instruments would also be available upon request.

Failing having a meditation room available, what can those who work in offices do to maintain their balance, relax, and be their most creative and awakened during a busy workday?

1. Make ample use of one-minute meditations, outlined in Chapter Four. Obviously, this will be easier if you are working in a private space than if you are sharing an office.

2. Practice the type of open-eyed alpha production described above. You don't need to be a visual artist to make use of creative states of consciousness with your eyes open.

3. *Breathe!* The most calming action you can take when faced with stress is to consciously focus on slowing your rate of breathing. For example, while you are listening to problems or complaints from a superior, you can at the same time be aware of your rhythmical and slow breathing and your relaxed heart rate. This not only helps to keep you calm but gives you the detachment that helps provide proper perspective when dealing with crises.

4. Also, when faced with stressful situations, make a complete energy circuit in your body. Sit with the palms of your hands together or—just as effective—the tips of the thumb pads and middle fingers touching one another. This helps contain the flow of energy within your body and maintain centeredness and balance.

5. Sit with your spine straight and relaxed, and your legs uncrossed. This also unblocks energy, which can then be called upon for use.

6. *Sensualize!* If you are facing a very difficult encounter or situation, take a few minutes to be by yourself before it begins. Using all of your senses, imagine the situation occurring in the most successful and healthy way possible. Imagine your own actions and reactions to be calm, strong, creative, and appropriate.

7. Use ordinary activity as a meditation practice. For example, when you are going to the watercooler for a drink, take the opportunity to be awake and aware. Be aware of each move as you make it and be very present in the actual act—not drawn back into the past or forward into the future. Be sensually aware of the smells, tastes, sights, sounds, textures, and kinesthesia of the situation. Savor every second of the experience, while remaining in the present.

8. Look for allies among your coworkers. You might be surprised to find other meditators more prevalent than you thought. There is support in numbers—if meditation becomes an acceptable and even pleasantly anticipated topic of conversation, your practices will be supported and you will feel freer to practice more frequently and more openly.

9. Support others in the need for and value of contemplative time.

BRAINWAVE MEASUREMENT AND TRAINING IN THE CORPORATE WORLD

Brainwave assessment has been used effectively in executive searches and corporate headhunting. Management consultants often have a variety of tools for determining the suitability of the candidates they are interviewing. These include written psychological testing, aptitude tests, IQ tests, and subjective interviews. When an EEG profile is added, it can potentially be the tiebreaker that could help determine which candidate is more suitable for a position. This can be especially effective when there is a language barrier on the written tests rendering these documents less dependable, especially for a position that does not require much use of language.

TAKING BRAINWAVE TRAINING INTO A CORPORATION

Corporations work on many levels. If you have limited time and resources, the most effective use of brainwave training within the corporation would be at the upper-management level. Helping to create awakened minds in the key positions within the corporation will have a positive and immediate effect on subordinates, even if they don't receive the training themselves.

When both time and financial allocations were limited, I have used the following model when working within the corporate setting. I have conducted individual brainwave training and executive development within the upper-management team and held workshops and group training for the middle-management personnel. This trickle-down theory is obviously not optimum, but the results were beneficial to the corporation as a whole.

Many leading executives I have measured come close to having an awakened mind brainwave pattern. The alpha, theta, and delta are readily accessible, but the individual often needs training to reduce splayed beta patterns. Once the awakened mind is developed, it is then necessary to stabilize the pattern so that it will be more consistently present.

I have worked in a number of executive development situations and addressed a variety of needs. For example, one corporation engaged me to do individualized training in brainwave development. The goal of this training was to improve their management team's access to the full range of their intuitive and intellectual capacities and so to enhance their decision-making abilities. Corollary benefits also included training in deep relaxation, stress management, and focused problem solving.

The CEO of this company wrote, "Executives who have participated in the program are universally pleased with the experience and the results. This assessment is particularly meaningful coming from a few who have acknowledged an early skepticism of the program . . . the Awakened Mind Program can be a valuable contributor to personal development and team performance."

Those who initially participate in this work purely for enhancements to their productivity in the corporate world are often startled and pleased by what one VP called "the value and inevitable focus on spirituality that evolved from the work." I often find that individuals who begin brainwave training for a specific, objective purpose also become quickly interested and involved in seeking higher levels of spiritual consciousness.

BRAINWAVES, CHILDREN, AND LEARNING
"What happened to my child?"

Children have the capacity to develop an awakened mind brainwave pattern easily and naturally, and do so well before school age. I firmly believe that this state is inherent but gets *trained out of us* at an early age. Common methods of teaching in our schools and the overcrowded situation in most of the nation's classrooms contribute to this loss. I have stated earlier that I believe the awakened mind brainwave state is a form of evolution. It appears early in a child's life, but then gradually disappears as the young person tries to fit into the accepted patterns and attitudes of society. I have generally seen the pattern start to disappear around the second or third grade.

"Don't daydream!" (get out of alpha). "Pay attention!" (with your beta waves). "Don't have those fantasies" (stop your theta waves). Most of the messages to the tender consciousness of a child trying to make sense of his or her experiences are aimed at increasing beta brainwaves and reducing alpha, theta, and delta brainwaves. To get along in our modern Western society, the child can be forced to squelch those states of consciousness that later he is trying so desperately to recapture. If we could understand and honor states of consciousness, both in our school systems and in our homes, we might be able to prevent the breaking down of the connections between our conscious and our unconscious minds.

It is fascinating to watch the brainwaves of a child develop in terms of consciousness. I first looked at my son, John's, brainwaves when he was two weeks old. He produced primarily delta and theta waves. Over the next months, I saw the addition of alpha and saw his mind struggling to add beta consciousness to the already strong unconscious and subconscious parts that were present. Dangling a toy in front of him or jingling some shiny keys would stimulate strong bursts of beta, but he would not retain those higher frequencies for any length of time. Finally his brainwaves stabilized, and by about age five he was producing a very strong awakened mind brainwave pattern. At eleven years old, he is still producing this pattern, but not continuously. He is more prone now to the splayed beta that is so prevalent in most adults. He is still able, however, to shut off the splayed beta and return to his awakened mind state when he wants to.

I taught John a form of meditation and guided imagery from a very early age, often as a substitute for storytelling. One of his favorite meditations was the Purple Planet.

THE PURPLE PLANET

Children tend to need shorter pauses. Experiment with your child to determine the most helpful timing.

Sit or lie down quietly and close your eyes.
Get very comfortable.
Let out a big sigh and breathe all the air out of your body.
Aahhh . . .
Now continue to breathe slowly and deeply.
Now go inside each part of your body, and feel how relaxed it can be.
Your head,
your neck,
your shoulders,
your arms . . . and hands,
your chest,
your stomach,
your whole back,
your legs . . . and feet.
Feel your whole body relax.
Try not to think about anything except your body feeling good.

Now go to the top of your head,
and imagine that you are inside it.
Now you go up and out of the top of your head . . .
straight up . . .
And you keep going up . . . and up . . . and up.
You go way up above the planet . . .
past the moon . . .
past the other planets . . .
way up into the stars.

Way in the distance you see a tiny purple planet.
And you get closer . . . and closer . . . and closer.
And the planet gets bigger . . . and bigger . . . and bigger.
You know that you are going to land on this planet and have wonderful
  experiences here . . .
When you land on the purple planet, you see purple sky and purple
  ground, and purple buildings . . . purple everything.
There is a friendly group of people from the purple planet there to meet
  you when you land.
You spend some time with these people and have fun.
And they have a lesson to teach you—something very important.
Spend some time now with these wise people, and learn what they have
  to show you.

(pause for up to one minute)

And when you are ready, you are going to leave the purple planet and
  come home.
Remember what the people there taught you, so you can think about it
  later.
And now you begin to say good-bye.
And you fly all the way back through the planets,
back past the moon,
back to the earth,
back into this room,
and back down through the top of your head.

When you are ready, you can stretch, breathe deeply, and wake up.

After taking my son through this meditation, I would ask him to tell me all about what happened on the purple planet and what he learned there. John had many different types of experiences, but he almost always had a lesson of some moral significance. He would report, "They taught me to be nice to all people," or, "Don't be mean to other people," or, "Share your things," or, "Don't tell lies." Sometimes they just showed him "neat stuff." The people on his purple planet were green with yellow polka dots, and they all looked like one another. John looked forward to going back there and would ask for that meditation before bedtime.

A CHILD'S USE OF MEDITATION

For a child, as for an adult, there are many different ways into the meditation state. Once there, the experience of inner peace, silence, and serenity can be the same for many people. But accessing this inner tranquility can be a different process for each individual.

My son calls meditation "deep, inner nonthinking." Once, while on a visit to Disneyland for his ninth birthday, John and I were on a very crowded bus. We were on the way home, and everyone was exhausted after a long and arduous day of fun. The voices began to escalate, and all of a sudden there was an incredible din on the bus. John complained irritably about the sound.

He then became very quiet for a while, and I thought he had gone to sleep. About half an hour later he stirred and said, "That was nice!" He had imagined that everyone from the bus was outside with him in an open field. He shouted loudly at all of them to "shut up!" They instantly became very quiet, and, he said, "You could hear a pin drop." He then sat in that silence, enjoying his inner peace. Meanwhile, the actual noise level on the bus had not diminished.

ATTENTION DEFICIT DISORDER AND LEARNING DISABILITIES

Brainwave training is now successfully being used for helping attention deficit disorder (ADD) and other related disorders, including

attention deficit hyperactive disorder, learning disabilities, opposi-
tional deficit disorder, conduct disorder, anxiety, and depression.
Many children and adults with these disabilities have an overabundance of theta and a deficiency of beta brainwaves. The generalized
protocol of brainwave training for such conditions is theta suppres-
sion and beta enhancement. A number of people have been working
successfully in this arena for several years.

Sometimes the things that stimulate our brainwaves and our creativity are the ones least expected, as the following story illustrates.

DEADHEAD

One of my teenage clients was classified as learning disabled. She was
also a Deadhead, one of the group of dedicated fans who would
follow the band the Grateful Dead around the country, attending
their concerts. She was still in high school, so she traveled only
during vacations.

A.D. disliked school and was having trouble with her grades, especially with her homework. She wanted to listen to the music of the
Grateful Dead while she studied and was adamant that this helped
rather than hindered her learning ability. Her parents, however,
steadfastly refused to let her listen to her music during homework
time.

Originally, when I heard her story, I tended to agree with her
parents. A.D. had an awakened mind brainwave pattern with lower
than average beta and higher than average theta. I was working with
her to increase her level of beta, training her to be more conscious of
how to turn it on. She brought homework to the sessions, and we
explored ways of increasing her beta retention while she was working. Little by little her grades improved and her D's gradually became
B's.

One day A.D. arrived for her session with a Grateful Dead cassette
tape she especially liked, and wanted me to listen to it. I agreed, but
only after I had hooked her up to the Mind Mirror, because I
wanted her to have a full session.

I saw her normal pattern on the screen until we began to listen to
the music. Then what I saw amazed me. Her beta increased dramati-

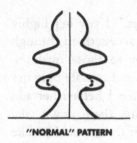

"NORMAL" PATTERN

LISTENING TO
GRATEFUL DEAD

FIGURE 5 ■ DEADHEAD

cally while she listened to the Grateful Dead (see Figure 5). I gave her some schoolwork to do, and she maintained the beta while working *and* listening to the music. I continued experimenting with her until I was satisfied that listening to her favorite music actually did improve her brainwave pattern for studying. The hard part was convincing her parents.

As in the case of A.D., I have noticed that children with learning disabilities can come close to having an awakened mind brainwave pattern, simply with less beta than is optimal. I prefer to work with them to enhance their beta, without necessarily suppressing their spiritual and creative theta waves, an action I fear would limit their innate genius and evolutionary abilities.

I would prefer not to think of normalizing these children, because I am not sure that our medical and societal standards of normal are necessarily optimum. Children with learning disabilities may actually be more gifted than is usually realized. It is up to us to make sure that those gifts are not routinely trained out of them for the sake of preserving normality. After all, if normal is always the goal, evolution cannot occur.

I would prefer, instead, to teach them to value and to use the gifts that they have. It is possible for them, with the right attitude and training, to turn their disability into an advantage. Inherent within this is the need to understand that the strong points of each person's brainwaves need to be sought out and developed to maximize his or her individual potential.

The teenager A.D. was unusually creative, bright, a good conversationalist, socially aware, and very intuitive and empathetic. As her father had noted, "She always lands on her feet." I saw her highly developed theta brainwaves and her ability to access them through strong alpha as the major contributing factor to those qualities. When I encouraged her to shut down her theta and activate only her weak beta, she became listless or agitated. When I helped her add beta to her alpha and theta instead of shutting them down, she seemed to come alive. I felt that taking away her theta would be like taking away her innate ability to land on her feet.

Ours is not the only culture where the school system tries to mold children into a prescribed model.

THE BASEBALL PLAYER

When I was in Taiwan, I was asked to work with the twelve-year-old child of a clothing manufacturer. His family lived in a village outside Taipei. They were well educated and wanted the best for their son. L. was the only boy in a family of girls. With several sisters and no brothers, he felt a lot of pressure to succeed. His family felt that he was having problems in school, and where he comes from that is a very serious matter. In his school there was a large sign in front that continuously posted the top three students *and* the bottom three students. Fear of failure and fear of disappointing his father was a major motivator.

I arrived in his home and was greeted by eight or nine members of his family. His elderly aunt and uncle had also been brought in to see me. The aunt was very ill, and the family asked if I would be willing to see her first. Recognizing her great need and the distance she had traveled, I agreed. While I was working with the aunt, the rest of the family prepared a huge feast for me. I was beginning to wonder how the session with this child was going to take place.

After dinner we assembled in the family room. I asked for privacy with the boy, saying I thought he would learn better. This caused an uproar and disagreement because *everyone* wanted to be present. They all thought it was their right, and they all thought they could help. This was perhaps the strangest individual session I had ever done.

L. did not speak the same dialect that my main translator spoke, so I needed to enlist the services of my backup translator, but he was not completely proficient either. As we proceeded, various family members would also offer their translations, sometimes arguing at length over the meaning of one of my words.

Eventually I understood that the boy's troubles all seemed to boil down to self-esteem, creativity, and sports. The child badly wanted to do well in school and in sports, but every time he expressed his desire, one of his older sisters or mother would say, "Oh, you could never do that!" I noticed this in different ways again and again.

His brainwaves were remarkable. He had rounded but strong beta, consistent and continuous alpha, and very strong consistent theta, with high-amplitude delta flares (see Figure 6). I felt that the huge amount of theta contained unfulfilled creativity that was continually being squashed.

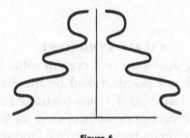

Figure 6

I took him into a meditation and told him that I was going to get him to tell me a story. With that, his family exclaimed, "Oh, he could never do that!" They all agreed he was not able to tell a story. After strongly asking them to be quiet, I instructed him to begin by creating the environment he would like to be in. He said that he saw himself on a baseball diamond and that he was up to bat. With this, there was another loud chorus of protests from his family, saying that he could *never* play baseball.

At that point, I insisted that they be quiet and let him try. What unfolded was a beautiful story about how he scored the winning point with a home run, leading his team to victory. He told it so intricately, delicately, and bravely, building the tension and expectations, because we truly didn't know, as we were listening, whether he was going to win or lose. When he finally won, he threw his arms up in victory, jumped off the couch, and yelled. His family burst into applause and gathered around hugging him.

Later they told me that this was the first "victory" that L. had ever experienced. They also said that they had not believed that he could possibly tell a story, because he had never done so before. L. fell in love with the ESR meter, attributing some magic power to it and clutching it to his chest. I asked his father if he would consider letting him play real baseball. His father was very doubtful, so I made a deal with him. I said I would give L. the ESR to keep if his father would let him play baseball. He agreed. After he had given me the prearranged payment for my work, in gratitude L.'s father took me into his clothing store next door, again with the whole family happily watching, and invited me to pick out my three favorite coats, clothing I have now long cherished.

REYE'S SYNDROME

The *gifts* of learning disabilities and even of some brain damage are often well disguised and overshadowed by the difficulties. A fourteen-year-old client I shall call T.J. had contracted Reye's syndrome, a rare disease of the central nervous system, in her preteen years. During her illness her brain became swollen, and she was near death for many hours. Both she and her mother felt that only her mother's vigilance and psychic insistence that she remain alive kept her living. From that time on, she and her mother were bonded to the point where each constantly and effortlessly knew what the other was thinking and feeling.

When T.J. came to me, she had a long list of complaints, including problems at school, social difficulty, fear, sadness, anger, and a feeling that she was closing herself off behind a brick wall. She also had a distorted sense of time and reality. T.J. had many odd and inexplicable experiences. Once, while in a house she had never visited before, she pointed to a wall and exclaimed, "There is supposed to be a fireplace here." There was indeed a fireplace there that had been walled up for many decades. While walking on a local sidewalk, she moved to sidestep a tree that wasn't there, only to find it had been cut down many years previously. She sometimes went into trances and saw people in eighteenth- or nineteenth-century dress in houses that stemmed from that era. She read minds, felt other people's feelings, and almost continually felt bombarded by the thoughts and emotions around her. She felt visited by spirits, some friendly and supportive, others not. Her mother, who was a student of mine in the college where I was teaching, eschewed any traditional medical treatment or drug therapy and brought her to me for brainwave assessment and possible help.

When I saw T.J., she had one brick left to go in the imaginary wall she was building around herself. Her brainwaves were primarily theta and delta (see Figure 7).

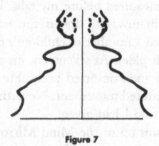

Figure 7

At first T.J. did not want to be there. I gave her the choice of leaving or staying, promising that I would explain her decision to her mother so she wouldn't get into trouble. Then I invited her to talk. Over the weeks she grew to love the sessions. She felt supported, heard, and understood. I showed her the vastness of her theta and delta waves and explained the nature of the subconscious to her. She did not want to lose her abilities; she just wanted to temper them and learn to get along in school and society.

Brick by brick, we took down the wall. I remember the day she finally said good-bye to the last of the spirits. Her theta was still strong, but she was learning to control it. She was also learning to deal with the subconscious content, the insights and awareness that continued to beset her. Her grades improved, and her participation at school increased. Her mother's anxiety about her, as well as the inappropriate bonding, lessened and finally disappeared. I taught her how to use her intuition and empathy but not be ruled or suffocated by them.

Now, almost ten years later, at twenty-five, she is happily married with a one-year-old child. She remembers those days but doesn't dwell on them, having the life now that she never dreamed she would be able to have.

A GENIUS IN A DAMAGED BODY

I was conducting a lecture/demonstration for a group of about eighty therapists from a center for disabled children in the South. They had agreed to provide me with a number of volunteers whom I had never met or measured before my talk. I had given my usual introduction about brainwaves, drawn the various patterns on a blackboard, and shown a number of children's patterns. The last boy I hooked up was a bit older. At seventeen, his cerebral palsy had left him with a very stiff and contorted body. He was unable to speak intelligibly and had limited movement. Nevertheless, he had listened attentively to the lecture I had given.

When I hooked him up to the Mind Mirror it showed a perfect awakened mind brainwave pattern. There was a gasp of astonishment from the group of therapists, and I said to him, "Do you realize you have the brainwave pattern many people want?" He nodded his head vigorously, and emotionally indicated the words, "I know, I know!" The look of exoneration on his face cannot be described. There was not a dry eye in the house. It was only later that I found out from his mother that this child was brilliant in certain areas. If you left him alone with a broken television, he would fix it. His understanding and knowledge of electrical wiring was at the level of a genius, yet he had been overlooked. His mother said he had just somehow slipped through the cracks. She was ecstatic to see her belief in her son reinforced by the Mind Mirror. Armed with information that could change his life and hers, she felt she had been vindicated. She believed now that the brilliance she had always known he had could indeed be recognized and developed.

As we have seen, the act of creating can take many forms. The brainwaves of creativity are applicable in almost every aspect of your life, if what you want is to originate, improve, develop, or respond better to life's challenges. Whether you access your creativity during your early morning time, in the shower, or during a meditation, seek to recognize and perpetuate your "ah-ha" experiences, or experiment with your mental fluency in the office, you will be working with the brainwaves that give you access to the deeper realms within you. With practice, creativity can soon become a natural part of your way of life.

1. C. Maxwell Cade and Nona Coxhead, *The Awakened Mind* (New York: Delacorte Press/Eleanor Friede, 1979), 58.

VII

BRAINWAVES
IN RELATIONSHIPS
EMPATHY, INTUITION, AND CONNECTION

Our brainwave patterns have a lot to do with how we experience and interact with the outside world. Individuals relate to others through the use of their brainwaves; therefore the brainwave pattern that they are producing helps shape the experience of their communication. In discussing our brainwave response to other people, I would first like to develop your understanding of *delta waves*. Delta brainwaves represent the unconscious mind.

Though usually thought of as the brainwaves of deep sleep, in a waking state delta waves are often referred to as the *orienting response*. Delta waves help orient us in time and space. On the very deepest level, they are our beacon that senses danger and safety. They serve as our automatic tracking device, our scanning function, and enhance our instincts, intuition, and ability to discriminate. This very primal, almost animalistic response can also be fine-tuned to sense emotions, needs, and attitudes in other people. It is often present when there is an unconscious connection or intuitive bonding between people.

I like to call delta a kind of *radar*. As such, these brainwaves are a major component of our intuition and empathy. Delta waves are often present in people who have a skill at reaching out to others in order to understand or help them. Psychotherapy and the healing arts tend to attract people with an abundance of waking delta. These very slow, low-frequency brainwaves are also present during most experiences of psychic phenomena and ESP.

PERSONAL EXPERIENCES OF DELTA

When are you likely to be producing higher amplitudes of waking delta waves? There are many times you may have had an inexplicable psychic knowledge on a very ordinary level:

- You *know* the phone is going to ring right before it does.
- You think of a friend a few minutes before you bump into him or her.
- You feel an unexplained pain, only to find later that your child was hurt at that exact time.
- You know what someone is going to say before they say it.
- You know *exactly* what someone is feeling, even though they are trying hard to mask it and other people are fooled.

Some people are capable of talking to one another with these brainwaves without opening their mouths or voicing a sound. A student of a well-known spiritual teacher writes, "We sat with each other for over an hour. We had a long and intricate conversation in which I asked him many personal questions, and he gave me the answers. It was only when we had finished and I said, 'Thank you,' that I realized that for the entire time neither one of us had uttered a word!"

BUT I DON'T LIKE MARMALADE

I lived for a time with a very close friend in London who was, and still is, what we might call psychic. We had much intuitive understanding of each other and many "silent conversations," but I will never forget the day she radared me to buy marmalade. I was out shopping with only a small list, a very quick job. While I was in the store, I stopped at the marmalade shelf and picked out an expensive, high-quality brand of the orange jam. Now this would not seem odd, except for the fact that I disliked marmalade and would never have consciously bought it. When I was at home unloading the groceries, Elizabeth breezed in, picked up the marmalade, and said, "Oh good, you heard me. Thanks!" All I replied was, "I *wondered* why I was buying that."

That was eighteen years ago. This same lady, Elizabeth St. John, subsequently became one of England's most accomplished healers. She feels the pains, problems, and traumas of others in her own body. She is then able to give insights to her patients about the nature and cause of their difficulties, make accurate diagnoses, and help them begin the transformations on an energetic and etheric level that ultimately can translate into the physical body as healing.

DEVELOPING DELTA BRAINWAVES

I have learned that trying to develop delta waves with direct biofeedback does not work well. When I have given feedback to subjects who have difficulty producing delta, their delta almost invariably disappears, even when it has been spontaneously strong. I have found that their delta often diminishes as soon as they are made consciously aware of it. Developing delta is a function more of learning a way of being than of taking an action to create a specific brainwave. With that said, what follows is an exercise that helps develop your radar and intuitive ability. This meditation is adapted from an exercise I originally did eighteen years ago with Max Cade.

THE BUBBLE

This meditation is best undertaken with a group of people, preferably sitting in a circle. Before you close your eyes, take a look around the circle to see where each person is sitting, and take a mental photograph of what you see.

Close your eyes and take a few moments to enter into a state of meditation.

Now begin to imagine yourself alone, seated at the center of a large bubble—like a soap bubble. This represents your consciousness, everything that you are aware of at this moment inside and outside yourself.

Now picture yourself at the center of a much larger bubble—the size of a large room. This bubble represents your preconscious, all the things that are not now in your mind but that you could recall instantly if you wish.

And now imagine yourself inside a still much larger bubble—the size of a football field. This represents your unconscious mind and contains everything that you have ever been aware of. You have the capacity for total recall of this material.

Now return again to the first bubble . . .

This is your personal space . . .
Feel it around you . . .
Just be aware of this bubble enclosing your consciousness . . .
You might even see it shimmer as it envelops you from the top of your head to the bottom of your feet . . .
your personal space . . .

Now, very gently and slowly, begin to allow your bubble to grow . . .
sacrificing your personal space to take in the person on one side of you . . .

What does it feel like when they enter your space? Try to be nonjudgmental, simply experiencing their energy as they enter into your space, and noticing the response within yourself.

Now allow your bubble to expand a little more to take in another person in the room. Again how does it feel? What are you experiencing?

Allow your consciousness bubble to continue to expand as you take in more people in the room. One by one you can feel them enter your bubble—your personal space. What does it feel like?

Don't analyze, interpret, or judge. Just experience.

Allow your bubble to grow larger and larger until it encompasses the whole room, until we are all enclosed in one large bubble—a group-consciousness bubble.

Just be aware of the feeling of this . . .

And soon you are going to begin to allow your bubble to shrink, reducing in size now so that it excludes first one person . . . and then another . . . and another . . .

Be aware of how you feel as each person leaves your bubble. What do you experience as each person pops out of your personal space? . . . and as the bubble continues to get smaller? . . .

Again, don't analyze, interpret, or judge. Just experience . . . .

. . . until, eventually, you are back in your own small bubble, your consciousness bubble, enclosing you from the top of your head to the bottom of your feet.

What does it feel like?
Experience your own consciousness around you . . .

Now take a few minutes to meditate on your experience and what it meant for you. Which was more comfortable—the large expanded bubble including the whole group? the small bubble of your own personal space? somewhere in between? What does this tell you about your boundaries? What have you learned about your needs, wants, and feelings?

Crystallize your experiences in two or three words or sentences. And when you are ready, allow yourself to begin to return, bringing this information back with you.

When you are ready, take several deep rapid breaths and allow your eyes to open, awakening alert and refreshed, back in the outside space.

RESPONSES TO THE BUBBLE

The responses to this meditation vary greatly. Some people experience a state of bliss at the end of the exercise and others a state of discomfort or anxiety. The following are a selection of individual responses:

It felt good to expand and be connected with other people—I enjoyed it. But it also felt good to return to the safety and privacy of my own space.

I couldn't stand letting anyone into my space. I felt claustrophobic and invaded!

When I returned to my own space, I felt a deep sense of loss and abandonment.

At first I didn't think I was going to like opening up to the other people. I felt like I was exposed and vulnerable. But then when you had us shrink our bubbles again, I didn't want to. When I was back in my own bubble by myself, I felt alone.

I found I could only let three people into my bubble at a time, me and two other people. It didn't matter which of you were in there, but if a new one came in, one had to go out. I am the same way with entertaining in my home. I will invite no more than two people over to visit me at a time.

I loved it. I didn't want to come back. It's a state I'm very used to and very comfortable in.

I didn't want to let certain people into my bubble. Others popped in right away. I found myself judging people as they came into my bubble. I learned something about my excessive judgment and also about my ability to be discriminatory about who I let into my personal space.

Each one of these could be taken further in an exploration of what the meaning of personal space has to the individual. Issues of safety, abuse, abandonment, and isolation are common, as are feelings of warmth, opening, compassion, and love.

As you can see, the responses are very varied. There are no right and wrong responses. This is a learning experience in which you can find out about your relationship to boundaries. Learning appropriate boundaries is the task, and the skill is being able to consciously shift how and with whom you connect. Learning expansiveness, openness, and nonjudgmental acceptance of whatever *is* accelerates your own development. But there is indeed a need for discernment about whom you let in to your psychic space, for how long, and for what purpose. There is also a need for careful discrimination about when it is appropriate to enter someone else's personal space.

I have often suggested that a student try this exercise outside of the group environment—with an individual who is not aware of what the student is doing. This is an especially helpful exercise to try with someone you are having trouble communicating with. This may be your spouse, your boss, a colleague, a waiter in a restaurant, or cashier at a store. Notice what happens to your communication after you have put them into your consciousness bubble.

THE PROBLEMS OF TOO MUCH DELTA
"I *always* know what everyone else is feeling—and it makes *me* feel terrible!"

Metaphorically, delta is both a radio transmitter and receiver. Some people transmit better; some receive better; some do both equally well. You can learn to regulate your transmitting and receiving—to turn it up or down. Imagine that you put it on a rheostat that allows you to control or determine the amount of information you send and receive on an unconscious or psychic level. Then when you do receive information, learn to detach from it.

One of the common problems of sensitive people is an excess of delta brainwaves. If you are one of these people, you might appear overly sensitive to the thoughts, feelings, and needs of other people. Your unconscious mind is picking up too many stimuli from the outside world and absorbing them inappropriately. What do you do with these stimuli? A common response is to assume that other people's painful emotions are your own, with very little differentiation between what is yours and what is not yours. Or, because you are more *aware of* other people's discomforts, you might feel a need to take responsibility for them—*to fix them*. Sometimes this desire to fix them is simply an urge to alleviate the pain that you are feeling yourself as a result of being too interconnected with other's emotions. Sometimes it is a more complex psychological game: because you can feel it so strongly, you must somehow have caused it. Therefore you feel *guilty* unless you do something about it.

This is the arena in which codependence rears its ugly head. The children of dysfunctional families, adult children of alcoholics, and childhood abuse survivors all have something in common—*hypervigilance*. This anxiety of excessive watchfulness stems from the need in childhood to be ever alert to the constantly changing and potentially dangerous experiences in an unstable home life.

High-amplitude delta waves have a number of implications for your relationship with the outside world. Experiencing them may cause you to

—feel bombarded by the emotions and needs of others
—experience the physical pain of others
—have a general feeling of being overwhelmed
—feel sensitive, vulnerable, and shaky in relationship to the outside
—feel overly responsible for the state of another
—feel guilty and burdened by responsibility for the condition of humanity in general

The answer may not necessarily be to turn down the volume of the delta waves, but rather to develop a healthy screening ability and proficiency in detachment. Some people with these responses feel the need to isolate themselves from the stimuli of the outside world. They may feel confusion and pain when they are exposed to this kind of input. If you can learn to expand yourself while you are receiving these stimuli, you will not need to reject them or defend yourself against them; nor do you have to take them on or integrate them into yourself. Rather, you experience the stimuli, acknowledge them, honor them, and let them *pass through you* without letting them stick or feeling responsible for fixing them.

This does not mean that you do not take a healthy and appropriate responsibility for your own part in relationships or for the health and well-being of the planet. It simply means that *you do not have to suffer* while you are using your discrimination to determine right action and appropriate responsible behavior.

CODEPENDENCE

Melody Beattie defines codependence in *Codependent No More* as follows: "A codependent person is one who has let another person's behavior affect him or her, and who is obsessed with controlling that person's behavior."[1] Among the many characteristics of codependency that she lists are

—feeling responsible for other people's wants, needs, and feelings
—trying to anticipate other people's needs in order to fill them
—feeling obliged to try to solve other people's problems
—saying yes when you really want to say no
—expecting others to do all of the above for you

DETACHMENT

Beattie considers detachment to be one of the basics of self-care:

Ideally, detachment is releasing, or detaching from, a person or problem *in love*. Detachment is based on the premise that each person is responsible for himself, that we can't solve problems that aren't ours to solve, and that worrying doesn't help. We allow people to be who they are. We give them the freedom to be responsible and to grow. And we give ourselves that same freedom. We live our own lives to the best of our ability. We strive to ascertain what it is we can change and what we cannot change. Then we stop trying to change things we can't. We learn the magical lesson that making the most of what we have turns it into more.[2]

The *Serenity Prayer* used by Alcoholics Anonymous expresses detachment in a way that millions of people in twelve-step programs all over the world can understand: "God, grant me the serenity to accept the things I cannot change, the courage to change the things I can, and the wisdom to know the difference."

These words hold special importance for people with high levels of delta waves. Delta can make you oversensitive to the energies within other people. Practicing detachment offers a healthy alternative for dealing with that awareness.

Swami Rama explains that with the help of *nonattachment* we can overcome the obstructions on the path of self-realization. He illuminates four ways of removing these obstacles:

First, by renouncing or giving up the object of attachment—which is often quite difficult for ordinary people.

Second, by learning to view the object of attachment as a means rather than an end in itself—which involves transforming attitudes, thereby changing bad circumstances into favorable ones.

Third, by learning to act skillfully and selflessly, "surrendering the fruits of . . . actions for the benefits of others."[3]

Fourth, by learning self-surrender: "one surrenders himself and all that he owns to the Lord and leads a life of freedom from all attachments."[4]

BRAINWAVES IN INTERPERSONAL
RELATIONSHIPS

Brainwaves and interpersonal relationships is a fascinating area of study that opens the door to a deeper understanding of human communication and connections. But the experience of *entraining* can occur in everyday life. Entrainment between two individuals is the reciprocal brainwave connection that occurs when people are in tune with each other. Conversely, as in the case of two people arguing, brainwaves may show the opposite pattern, each person's being completely different from the other's.

Many couples I have measured have a tendency for their brainwave patterns to resemble, but not entirely mimic, each other's, especially those who have been together for a long time. For example, both partners may have strong alpha, although one's amplitude usually surpasses the other's. It is also possible that neither one has strong alpha, beta being the dominant frequency for both. If the relationship is strong, they may both have access to high-amplitude delta as well. However, these similarities are not a rule. Some of the best relationships are founded on opposites, and this includes opposite brainwaves. Lack of entrainment is *not* necessarily an indication of incompatibility. Jack Sprat and his wife worked very well together, as can the beta dominant and alpha dominant couple.

On the other hand, extreme opposite patterns can also be a reflection of irreconcilable differences. A couple locked into incessant arguing may not realize that part of their problem is the different natures of their states of consciousness. I have found that using EEGs while counseling couples can be a very useful practice, enabling me to point out through biofeedback what is happening to each partner's brainwave pattern during their interactions.

Figure 1 shows an example of a couple in deep argument. There are many ways to argue. I picked this one to illustrate because the interrelationship of the brainwaves is so graphic. In a resting state their patterns are different, but not abnormally so. In a state of arguing, you can see that, while he shoots out his beta waves at her, she

1A ▪ MAN RESTING

1B ▪ WOMAN RESTING

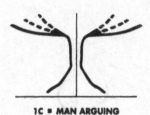

1C ▪ MAN ARGUING

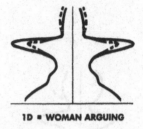

1D ▪ WOMAN ARGUING

FIGURE 1

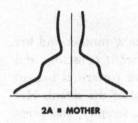

2A ▪ MOTHER

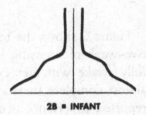

2B ▪ INFANT

FIGURE 2 ▪ NURSING

contracts her beta and shoots out her alpha at him. He is yelling, "I think . . . ," and she is crying, "I feel . . ." The visual effect on the Mind Mirror is astounding. You can actually see the brainwaves fighting! In a case like this I have been able to give the individuals feedback as to what their brainwaves are doing, and encourage them to learn some self-mastery of their own internal states as a useful tempering of their full-blown argumentative style.

3A ▪ HUSBAND

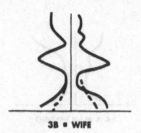

3B ▪ WIFE

FIGURE 3 ▪ LONG-TERM MARITAL PARTNERS, BOTH IN MEDITATION

4A ▪ MOTHER

4B ▪ ADULT DAUGHTER

FIGURE 4 ▪ MOTHER AND ADULT DAUGHTER IN MEDITATION TOGETHER

Figure 2 shows the brainwave patterns of a new mother and her two-week-old nursing infant. While the mother is conscious and fully awake with her eyes open, her brainwave pattern is one of almost complete theta and delta, mirroring that of the infant. Her reported experience is one of deep satisfaction, serenity, and overwhelming love and connection with her child.

Figure 3 depicts long-term marital partners. This particular couple has been married over forty years, the only marriage for both; they have lived in three countries, raised two children, and have a committed and loving life together. While the brainwave patterns are not completely entrained, they are very similar in their strong, consistent alpha and high-amplitude right hemisphere delta. When they came to me for marriage counseling, I found that they each needed to work on themselves individually rather than on the relationship.

When they both began sorting out their own issues, their relationship began to flourish.

Figure 4 shows the brainwave patterns of a mother and her adult daughter in meditation together. While they are not in exact entrainment and not trying to be, you can note the similarity of the two patterns.

CAN ENTRAINMENT BE A CONSCIOUS CHOICE?

So far, we have looked at entrainments in marriage, lovers, and families. These all appear to happen naturally without conscious intent. If you wish to experiment with consciously entraining with a friend (or with someone with whom you are having difficulty) and have no biofeedback system available to tell whether it is happening or not, you may wish to try meditating together. Pick a topic or method of meditation that you can both practice. Try synchronizing your breathing. Share the content of your experience afterwards. Practice the Bubble meditation and again share your experience. The more you can practice sharing your lower frequency and subconscious states together, the more likely you will be to share common brainwaves.

Conversely, what if you feel yourself entraining with someone and you *don't want to*? There may be times when another person would like to connect with you on a brainwave level in a way that is inappropriate or undesirable for you. While you probably will not have the biofeedback devices necessary to show you *exactly* what is happening, you can still feel the intrusion of unwanted intimacy. You might find it helpful at this point to return to your consciousness bubble and *exclude all other consciousnesses but your own*. In other words, thicken the skin of your bubble. It may also be helpful to consciously form a stronger connection with your own higher self or higher power—whatever form that takes for you. This alliance can be helpful in giving you the strength to ward off undesired approaches. Finally, you can also try shifting your own consciousness dramatically from state to state to effectively lose your stalker. Access high beta; reduce it and let it go; access strong alpha; move from the alpha back to the beta, and so forth. Keep moving around so that you don't settle on one particular state of consciousness that can be picked up.

HOW CONSCIOUS BRAINWAVE CHANGES MIGHT ALTER YOUR COMMUNICATION WITH OTHERS IN A POSITIVE WAY

There are times when you might find your brainwave pattern inharmonious with the situation you are in. By changing your state of consciousness, you can often change your experience of a difficult situation, allowing your response to be more positive.

ANGER

You are in an altercation. You begin to feel out of control.

Stop! Take a few moments to *reduce your arousal* by relaxing through deep breathing. Relax your tongue to reduce your beta waves. Internalize and focus on an inner center. If you feel safe doing so, bring your adversary into your consciousness bubble. Then begin your communication again.

FEAR

You feel paralysed. Your heart is racing with anxiety and you can't think what to do or say.

Again, *relax*. Close your eyes, even if only very briefly, to access some alpha waves. Quickly sensualize the type of communication you would like to be having. Take a few deep strong breaths, open your eyes, and begin again.

BOREDOM

You are trying to listen and stay awake, but you find yourself falling asleep or zoning out. You know that delta is just about all you've got.

Arouse yourself. Hyperventilate by breathing deeply and rapidly, for a few moments if necessary, to decrease your skin resistance. Sometimes this is all it takes. If you need more stimulation, think about something completely different while *keeping your eyes open.* Pick something that is exciting and stimulating for you. Feel the shift in your state of consciousness and *immediately* return your focus where it should be. Listen with attention, interest, and alertness to detail.

INTERSPECIES COMMUNICATION

By now brainwave interrelationships between people may seem like an obvious occurrence. Less understood, but very interesting, is the fact that human beings and animals may also have brainwave interrelationships.

The work I have done with interspecies communication and brainwaves involves horses and their riders or trainers. I fell into this work by accident when I met Linda Tellington-Jones, an internationally renowned animal expert, who developed the Tellington TTouch and founded TTeam, the Tellington-Jones Equine Awareness Method. Linda is also involved in interspecies communication. The first time she came to me for a brainwave profile, I monitored her while she was working on people. In this state she produced a form of awakened mind brainwave pattern that was heavily weighted with theta brainwaves.

I was interested to know if her students had a similar pattern, so we set up a test during one of her workshops at a Colorado ranch. I observed that all of the students who had studied the TTouch over a period of time tended to have strong theta and delta brainwaves in a normal resting waking state. Six out of the eleven people I measured had near awakened mind patterns in the left hemisphere, and one person had an awakened mind as coherent as Tellington-Jones's.

The Tellington TTouch involves a special way of pushing in a circular clockwise motion the skin of the horse, other animal, or human. Tellington-Jones told me that although it is hard to give a comprehensive definition of her work, her official description is as follows:

> The Tellington TTouch is a specific system of finger and hand touch that finds areas in an animal's body that indicate tension, fear, soreness or discomfort that limit its ability to co-operate or perform. Specific touches and movements are taught to reduce tension and to release discomfort and pain or the fear of pain. The touch has been used to speed up healing and to speed up recovery from injury and illness in animals including humans. Attitude and behavior also can be changed with the touch. It improves the ability to think and perform in animals. In humans it improves the ability to focus.

Our next step was obviously to attempt to monitor horses' brainwaves and then to see if we could observe any effect from the TTouch. We found that the basic resting state of the horses was primarily theta and delta with occasional flares of alpha. When the TTouch was administered we got an activation of all four categories of brainwaves on the horses. We saw that alpha especially was consistently activated during the TTouch, as well as some beta.

Later we used a remote telemetry unit to measure the brainwaves of the horses while they were walking. We saw activation of the higher frequencies in certain patterns of movement and not in others. I then simultaneously monitored the brainwaves of Tellington-Jones and a horse she was working on, and found a high level of entrainment occurring between the horse and the trainer!

THE PSYCHIC HORSE

Perhaps the most startling experience that we had took place while working with a two-year-old thoroughbred mare that the owner thought was crazy. Initially, this horse had highly scattered brainwaves and out-of-control, high-amplitude flares. She had exceedingly strong theta and delta and not as much alpha and beta as we thought there should be, according to the other horses' brainwaves. Tellington-Jones then spent some time doing the TTouch on her.

Afterward I was standing in front of a group of people talking about our discoveries and discussing this particular horse's difficulties. I explained that this horse could produce only theta and delta and was unable to produce alpha—whereupon the horse immediately produced strong alpha. When everyone laughed, I said, "O.K., but she still can't produce beta." When she immediately produced beta, no one laughed, because our mouths were all open! Time prevented us from experimenting further with this particular horse. I still wonder what would have happened if I had said "O.K., but she can't produce an awakened mind."

At the end of our investigation, Linda Tellington-Jones expressed a strong desire to continue our research with brainwaves and interspecies communication. She said, "To me this is just a start. I believe what we are going to find out is that when we touch an animal in this way, whether it be a horse, or a cat, or a dog, that we are not only affecting the health of that animal, but we are also affecting our own health. That animal becomes our teacher, and we are involved in that nonverbal interspecies communication which I think is going to make a difference to our planet as we learn to deal with each other that way."

THERAPEUTIC RIDING

Some time later I had the opportunity to again monitor horses' brainwaves, this time with children as riders, at a therapeutic riding center. I saw the same characteristic patterns of horse brainwaves as before; but monitoring even closer, I began to notice the difference in the organization of the patterns. The sprightly young frisky horse had a very organized pattern. The old nag had a disorganized pattern.

The children were disabled with a variety of conditions, including cerebral palsy and autism. The most disabled child was paired with the old nag because it was thought that its age and slowness would make it a more dependable partner. We found, however, that when the child rode the old horse with disorganized brainwaves, his brainwaves became more disorganized as well! When the same child rode the horse with the most organized brainwaves, his brainwaves improved and became more organized! Here we saw that the brainwaves of the horse had a definite impact upon the child riding him. We were able to learn to pick the most suitable horse for a child by using brainwave measurement. Clearly there is a lot of exciting information to be learned about how animals communicate with and affect human beings.

We are all social beings whose daily interactions cross a wide spectrum, ranging from our relationships with animals to interrelating with our partners, children, colleagues, or friends to interacting with the grocer, mail carrier, or banker to mingling with strangers in a crowd. We are constantly in some form of association with others. How we form and experience these connections is determined by and reflects our brainwave patterns.

By becoming intentionally aware of our states of consciousness as we interact, learning to utilize our radar appropriately, and practicing conscious connection, we can enhance our relationships and improve rapport. When we understand not only how our brainwaves are affected by others but also how we can affect other people's brainwaves, we increase our empathy, our communication skills, and our satisfaction in relationships.

1. Melody Beattie, *Codependent No More* (New York: Harper & Row, 1987), 31.
2. Ibid., 51.
3. Swami Rama, *Living with the Himalayan Masters* (Honesdale, Pa.: Himalayan International Institute of Yoga Science and Philosophy, 1978), 175.
4. Ibid.

VIII

FULL CIRCLE — CONCLUSION AND COMMENCEMENT

Fifty years from now, a book of this nature may be commonplace. But even now it is important that we reach into the future and see where we are going. The journey down the path of evolving consciousness is an exciting and fulfilling one. If approached in the right way, it can be fun, safe, grounded, and exceptionally useful.

Awakening the mind and experiencing higher states of consciousness is surely an evolutionary step. However, evolution is not simply a smooth and easy, linear developmental process. It has hills and valleys, surges and periods of dormancy, anticipated events and unexpected ones.

In this process, you are the one who counts. Your experiences, needs, attitudes, and approaches to life combine to create who you are. By gaining greater mastery over your own internal states, you play a stronger hand in your own destiny. Evolving your own consciousness takes you on a personal passage of transforming pain into joy.

YOUR PERSONAL JOURNEY

Once you have taken the first step on your path of personal evolution, the path becomes your own. You might walk parallel to others for a time, but no one else has your exact path. This experience can be exciting and unsettling at times, as if you are riding a small raft over large and seemingly endless rapids. At other times you may wonder what happened to your forward momentum as you sense yourself becoming stationary or even stagnant.

CHANGE

As you begin to gain self-mastery and awaken parts of yourself that may have been long dormant, changes may occur in your self-awareness and understanding. Your personal needs and desires, your attitudes and interests, your energy level and health, your relationship to your environment, your communication styles and interpersonal relationships, and your long-term goals and aspirations may become transformed in a number of interesting and surprising ways. There may be periods when you think that something is wrong with you, that you ought to be as you used to be, or more like other people. As you learn the language of higher states of consciousness, there may seem to be fewer and fewer people you can actually communicate with about your experiences.

As these changes occur within you, the personal transformation that you are undergoing will certainly affect your external environment and relationships and your internal experience. It is possible to go into denial that you are changing or to try to hide it from those that are closest to you for fear of negative repercussions. If you find this happening to you, it is time to seek companionship and guidance from others who are in the process of personal or spiritual evolution. If you look for it, a large network of people exists that can provide a support system. This might also be the time when a spiritual teacher appears.

WHAT DO I DO IF I AM
CHANGING AND MY SPOUSE/PARTNER/
FRIEND ISN'T?

This is a common dilemma with no absolute answers, because of the individual nature of each person's circumstances. There are, however, some generic experiences we all share. It's best not to try to cover up the changes or hide them from those close to you. Share them with an expectation that your nearest and dearest will be interested—either because of his or her own personal call for development, or out of genuine love for you and his or her desire for your own best interest.

But don't force your process on anyone. Some people, even those who love you, may not want to hear what you are going through. Or they may not be able to accept it because it pushes their own buttons. This is not a reflection on you! This is their own issue that they must deal with and that you cannot solve for them. All you can do is to back off and give them space to evolve in their own way in their own time—not in your way in your time. Nonjudgmental allowing and detachment are the key ideas here.

If you are developing states of consciousness that are positive, powerful, creative, stimulating, or pleasurable, you may even encounter jealousy if you try to share them with the wrong person or in the wrong way. Again, you have nothing to hide and nothing to be ashamed of. It's not possible for everyone to understand you and condone your personal process. Personal growth and evolution have caused many a parting of the ways. If your relationship is strong and there is a true desire to stay connected, your change and growth should not be a problem as long as you each maintain the independence, respect, and self-respect that personal development demands. Joint counseling to clarify needs, experiences, and expectations may be useful. Taking enough time and space to keep the communication flowing between you is also helpful.

Each case is slightly different. I have known many couples whose individual interest in developing consciousness and spiritual awareness has brought them closer together. One couple found that, while the differences in their personal habits, behavior, and emotional needs had tended to erode the bonding in the relationship over a twenty-five-year period, when they prayed together and focused on God their difficulties melted away, leaving them feeling strong and bonded. When one partner felt particularly distant from or disturbed by the behavior of the other partner, he would simply pray for her and surround her in light. On the other hand, when either of them tried to *control* the other's actions, needs, or feelings, the relationship would weaken.

Another couple with a five-year relationship has very different spiritual practices. She arises between 4 and 5 A.M. to begin a two- to three-hour practice of meditation followed by one or two hours of physical exercise. She had been following this routine for many years before they met. She also participates in several retreats every year and is continually reading spiritual or growth-oriented books. Her partner came from the corporate world and was not a meditator before they met. He soon realized that if he wanted a healthy relationship, he would need to let go of any idea of trying to break her of her habits. The more he accepts her needs, the more the relationship thrives.

She has learned, on the other hand, that if she wants to maintain a healthy relationship, she needs to devote time, energy, and presence to the connection. She has tempered her isolation, solo lifestyle, and withdrawal into meditation for the sake of developing a strong and healthy bond with her partner. They have each carried out changes with awareness and open communication. I cannot pretend that their relationship is always smooth sailing, but they have faced the challenge of merging two different spiritual natures by using the tools of compassion for each other, commitment to continue to fulfill their individual needs, and a dedication to the relationship that has manifested in a lasting and successful bonding.

Examples like these, however, are not always the case. Many couples do grow apart for diverse reasons, not just the emotional or spiritual growth of one partner. If separation is unavoidable, the parties involved can still attempt to remain true to their highest purposes and the best interest of everyone. In these sometimes complex and painful situations, meditation can provide a helpful solace.

RIDING THE RAPIDS

There may be particular times of tumult in your evolutionary process. Always remember, *keep breathing!* Don't be afraid to elicit outside help with your emotional, energetic, or spiritual process. *Just make sure that the individual from whom you are asking for help is sufficiently ahead of you on his or her own path to help you with yours.* Use your intuition to find whom you really connect with and trust. If you feel a red flag in relationship to another person, listen to your inner sense, no matter how experienced the individual purports to be.

I found my own teachers in many different ways. Two of them came to me through the following process. While I was living in London, Max Cade once guided me on a meditation that was intended to help me find the next step on my spiritual path. I have no memory now of his words, but I remember the vivid sensation of lying down on a stone slab, deep within a temple, surrounded by about a dozen robed and hooded figures. The majority of the experience was kinesthetic rather than visual. I felt the energy of what seemed like dozens of hands streaming into my body, gently rocking and swaying me. It was an exciting meditation, but as I found no immediate meaning in it, I simply filed it away in my memory and forgot about it.

The next week I was a principal presenter at a conference in northern England on self-healing. The week after that, I went for my monthly visit to Major Bruce MacManaway's healing clinic. One weekend a month he organized six or eight healers working in different rooms in a large house in Victoria, a fashionable old section of London. Using his pendulum, he would determine when you arrived which healer would be best suited for you.

I had no particular problem I wanted to work on that day, only the desire to receive positive energy. Aside from that, my back hurt a little bit from a dance injury. Bruce sent me to an upstairs room, where a man I had never seen before was waiting. I introduced myself and lay down on the table as instructed. I closed my eyes and, within minutes, was back in the meditation I had experienced with Max. As this man put his hands on my body, I felt the dozens of hands of the robed monks who had surrounded me.

After the healing I said, "I've felt this before. I know I'm supposed to be here." He replied, "I knew you were coming. I was at the healing conference last week and heard you speak. I've been expecting you ever since." I saw him once a week for the next year. He was the person who so helpfully guided me through the physical and energetic manifestations of my kundalini experience.

After several months of seeing him, when he knew about my work with brainwaves and meditation, he lent me a book called *Meditations from the Tantras* by Swami Satyananda Saraswati, saying he thought that I would relate to these teachings. Although I glanced at the book and thought it looked interesting, I didn't read it for six months. I simply carried it around with me. Although I had never heard of the author before, I knew that he was a powerful teacher and that the book carried important meaning for me. I felt safe when I had it and as if I was missing something when I didn't. When my healer friend finally asked me to return the book, I quickly skimmed through it, finding it very familiar, as if I already knew the material. (I have since acquired many of his books, and they still maintain the same power for me today after eighteen years.) Within a few months, I met Swami Satyananda in London on a visit from India and was able to work with him personally. Years later, I received visits from other swamis from his Bihar School of Yoga, who taught meditation in my center in Boulder.

Teachers may come in many forms. The next person to help you along your path might be your therapist, a friend or acquaintance, your hairdresser, or even a stranger. My son has often provided great teaching for me.

Once I found a teacher in a man whose name I never knew and with whom I had only a minute's contact. I was in Germany, preparing for a large audience-participation performance by my dance troupe. We had rented a huge hall and were setting up with less than an hour to go, when one of the vital pieces of electrical equipment went dead. I was somewhat in a panic and sought out the manager for help.

I approached him saying, "We have a real problem, and I don't know *what* to do!" He put a calming hand on my shoulder, looked me straight in the eye, and said with a heavy German accent, "Vee don't haff problems heer. Vee only haff situations. Now, vat ist your situation?" His voice is still in my head years later whenever a problem occurs.

If no teacher is available or appropriate, you may find help in literature. There are many books that may relate to individual and specific aspects of your own developmental process. An afternoon spent perusing the library or bookstores can be a rewarding pastime.

When the rapids are really intense, you can always rely on the dependability of change. Things *will* move forward! If you haven't already, now is the time to turn to and to lean on your own higher power—or God as you understand God.

SURRENDER

Sometimes it's necessary to *just let go*. Give up trying to control, and turn it over to God. Complete surrender to a higher power may be the only thing helpful when the rapids are really rough.

Alcoholics Anonymous is an organization founded in 1935 to help alcoholics deal with the disease of alcoholism. It is a spiritual program based on twelve steps the founders took. Many twelve-step programs have developed out of AA's original premise, and these reach out and help people in all walks of life. The first three of these twelve steps are particularly relevant to the spiritual process we are dealing with. They have helped many people to just let go.

1. We admitted we were powerless over alcohol [food, gambling, anxiety, people, places, things]—that our lives had become unmanageable.
2. Came to believe that a Power greater than ourselves could restore us to sanity.

3. Made a decision to turn our will and our lives over to the care of God as *we understood Him.*

Within these three simple phrases lies the essence of the wisdom of surrender—acceptance of one's powerlessness, belief in a higher power, and surrender to that higher power. These simple principles have saved countless lives and restored them to equilibrium and serenity.

The dichotomy of self-control and surrender is often hard for people to grasp. How can we evolve if, to do so, we must first learn to control ourselves, yet we are powerless over ourselves? How can both self-determination and powerlessness exist at the same time? This is almost like a Zen koan—"What is the sound of one hand clapping?"

Accepting your powerlessness and surrendering to the guidance and light of a higher power allows you to receive gifts of guidance and equanimity. These gifts give you the power of self-mastery. Therefore, through surrender to spirituality true power is experienced.

> The Tao of heaven does not strive, and yet succeeds.
> It does not speak, and yet responds well.
> It does not ask, and yet is answered with all its needs.
> It seems at ease within, and yet it follows a plan.
> —Lao Tzu, *Tao Te Ching,* chapter 73

SPIRITUALITY

When one is walking a path of personal growth, spiritual development naturally follows. Regardless of what religion, creed, dogma, or belief system you embrace, as you develop your brain to a fuller capacity, you will expand your spiritual awareness. As your unconscious and subconscious become more available to your conscious mind, a number of changes tend to occur. It is often through that subconscious and unconscious realm that we make our strongest spiritual connection. A leap of faith occurs that allows us to fully and truly believe in some form of divine order of the universe, some form of power greater than ourselves.

Though this God may be different from the God of our childhood, it will be a loving power, not a punishing one. It is a healing power that can be totally trusted and relied on.

It doesn't matter what form this power takes. Higher power comes in many names with many forms: God, Great Spirit, Buddha, Jesus, the Great Mother, Lords Brahma, Vishnu, and Shiva, Mohammed, Nature, Love, Light, and Divine Essence, to name but a few. Many people experience God inside themselves instead of outside themselves, feeling God as an inner light or a higher self. God can also take the form of a guru (*gu* = "darkness" and *ru* = "dispeller"), or teacher, a medium through which you are able to reach the higher essence. Sri Ramakrishna teaches, "But, in fact, God the Absolute and God the Creator are one and the same Being. The Absolute Existence-Intelligence-Bliss is the All-knowing, All-intelligent and All-blissful Mother of the universe."[1]

The following meditation from the Christian tradition, adapted by Anthony de Mello, is designed to help you to open yourself to the endless possibilities of God. It is inspired by the Hindu practice of reciting the thousand names of God.

"I propose that you now invent a thousand names for Jesus. Imitate the psalmist who is not satisfied with the usual names of God like Lord, Savior, King, but, with the creativity that comes from a heart full of love, invents fresh names for God. *You are my rock,* he will say, *my shield, my fortress, my delight, my song . . ."*[2]

Ultimately all forms of God lead to the same place—a place of absolute bliss, peace, and well-being. This state is one that words do not define—only limit.

I wish I knew whom to give credit to for the following meditation. I understand that it was found in Australia, written in Sanskrit on a scrap of paper stuck in a book by Swami Satyananda Saraswati. I was never able to find out who translated it into English.

PURE SPIRIT

I know that I am pure spirit, that I always have been, and that I always will be. There is inside me a place of confidence and quietness and security where all things are known and understood. This is the universal mind, God, of which I am a part and which responds to me as I ask of it.

This universal mind or consciousness knows the answers to all of my problems, and even now the answers are speeding their way to me. I needn't struggle for them. I needn't worry or strive for them. When the time comes, the answers will be there.

I give my problems to this great universal consciousness; I let go of them confident that the correct answers will return to me when they are needed. Through the great law of attraction, everything in life that I need for my work and fulfillment will come to me. It is not necessary that I strain about this, only believe. For in the strength of my belief, my faith will make it so.

I know that my body is a manifestation of pure spirit and that spirit is perfect; therefore my body is perfect also. I enjoy life, for each day brings a constant demonstration of the power and wonder of the universe and myself. I am confident . . . I am serene . . . I am secure.

No matter what obstacle or undesirable circumstance crosses my path, I refuse to accept it. There can be no obstacle or undesirable circumstance to the consciousness that is in me and serves me now.

SELF-MASTERY

The major theme of this book is self-mastery—the controlling of internal states, the ability to identify which state is appropriate at any given time and to enter it at will.

The impact that meditation and brainwave training might have on your life and your spirituality is profound, but it doesn't happen all at once. It is a process of evolution. The awakening of the mind has many facets and takes place in many different time frames in a wide variety of manifestations. Yet in the end, there is only *one* awakening. The unconscious becoming conscious is perhaps the most succinct way of describing the process.

Christopher Hills writes, "The evolutionary force is not interested in any particular philosophy, religious tradition, or scientific method, but in more real ways of self-mastery . . . The tools of religion and science in the evolutionary sense are no greater than their users."[3]

Swami Rama relates evolution to self-mastery by explaining that from the yoga philosophy of evolution, the very *act* of learning to control brainwave states leads to the process of evolution:

. . . growth is always based on the attainment of some degree of disentanglement from attachments, which allows one to observe something about himself and his world to which he was previously blind. This is the expansion of awareness. As awareness grows, one inevitably discovers within his new definition of himself a new ability to control. What he was previously blind to and controlled by is now within his power to regulate. Increased capacity for observation leads to increased capacity to control . . . As parts of oneself which were previously operating outside awareness are brought under conscious control, they gradually come to be more coordinated . . . What previously gave rise to conflict is now brought together in a harmonious whole. The ability to control results in a new degree of synthesis . . . This process of increased observation leading to increased control and increased synthesis is what makes up each step in the long journey of evolution.[4]

A VISION OF THE FUTURE

I have the vision of a future where the minds of men and women are awakened, where unconscious communication is customary, where synchronicity is expected, where healers work in hospitals, where schools teach brainwave training, where corporations and government agencies regularly use meditation, and where consciousness has expanded to include the soul and the spirit.

Someday, God willing, we will, as a collective and in our communities and families, honor, respect, and support the evolutionary changes occurring in higher states of consciousness. We will not view the changes in our brains as abnormalities. Instead, we will support the journey of awakening with research, theoretical training, and experiential instruction.

Counseling will be geared both to accommodate and encourage development of conscious awareness. Doctors will recognize the symptoms of a spiritual crisis or an unexpected and abrupt awakening and be trained to advise and support individuals rather than discount, discourage, or incarcerate them. Religions will truly explore their common ground in spirituality rather than cling to their own specific dogmas for self-identification.

Imagine, for a few moments, a world where all beings are focused on and supportive of awakening to higher states of consciousness. What would happen to poverty, to hunger and homelessness? What would happen to political unrest and war?

What would happen to stress and anxiety levels? What would happen to relationships, to family and community? What would happen to creativity? What developments would take place in the sciences and the arts? What would happen to religion?

Higher states of consciousness will not solve all our problems or create an instant utopia. The human race has to solve the earth's problems in a practical, realistic, and useful way; but imagine if the people intent on these tasks were awakened to their highest states of creativity and aware of their highest states of spirituality. What a very different world it would be!

I have hope for the future. I already see society around me changing in many slow, steady ways. There is an increasing consciousness of hunger and action to help the homeless. There is a growing holistic health care industry, and what was once "alternative" medicine is now considered "complementary" medicine. There is a developing environmental-protection movement.

I see a growing unacceptance of sexual harassment and discrimination, and an attempt at balancing male and female energies as men and women try to integrate a new understanding of gender. I see the struggle with racial discrimination slowly and painfully bearing fruit.

I see an expanding spiritual awareness that accepts increasingly diverse paths to a higher understanding. Yet at the same time, I see a greater integration and recognition that all paths lead ultimately to the same source. I see an openness for and a pursuit of knowledge that honors and even yearns for previously discounted spiritual traditions, such as Native American. I see also an increasing number of people attempting to live healthier lifestyles through correct diet, clear thought, and right action.

I am by no means assured of the ultimate success of these changes, but I have hope. It takes only a threshold to change society, not one hundred percent. Regardless of our degree of optimism or pessimism about the future, our primary responsibility is to change ourselves, and that goal is certainly attainable.

When we begin to heal ourselves, we begin to heal our families, our society, and our world. Through opening to this evolution and illumination, we can begin to make a difference, to ourselves as individuals and to others.

OPENING TO CLARITY

I would like to end the book with one last meditation. You can practice this as you would any of the previous meditations and imagery exercises in the book, beginning with sitting or lying comfortably and allowing your eyes to close . . . .

—Let yourself enter into your meditation state, into your relaxation, into yourself . . . Find the point of stillness and the point of relaxation . . .

—Allow the energy to flow in your body, in your chakras . . . to be open, to be clear . . .

—Allow yourself to expand . . . Allow yourself to reach up and become part of your higher self . . .

—And from this perspective, this expansion, feel what it feels like to be in your own Tao, on the correct path . . .

—And allow yourself to be permeated with clarity . . . Experience clarity in your mind . . . in your emotions . . . in your body . . . in your energy system . . . in your spirit.

—Experience openness, and in that openness, open yourself to clarity . . .

—And from this place of clarity and expansion, consider what it is that you need. It may be something small; it may be something large; it may be simply the ability to continue on your path in the way that you are doing . . .

—Just consider what you need, and reflect what it represents for you, what it means to you . . .

—Is there a major stumbling block to reaching your goal? . . . What needs to happen for you to get what you need? . . .

—And once again experience clarity . . . and feel yourself going even deeper, to a place where you can truly open yourself to the light . . . to your own connection with the divine, to your own awakening . . . to your own illumination . . .

—You can experience, or feel, the light coming into you from all around you. You can feel it coming in through your crown, through your third eye . . . through your throat . . . through your heart . . . your solar plexus, your navel . . . your second chakra . . . your root chakra . . .

—You can feel yourself taking light in through the soles of your feet, the palms of your hands, the backs of your knees, your elbows, the fronts of your shoulders, your ankles . . .

—You can feel light entering every pore of your body . . . and you receive it, almost as if you drink it in . . . until you are full . . .

—And even when you are full of light, you can feel it surrounding you, all around you, completely covering you . . .

—And finally you can begin to feel it emanating from you. Even as you are full of light, you are also sending out light, so you can feel light emanating from your navel, your solar plexus, your heart, your throat, your third eye, and your crown . . .

—light pouring out of you from every cell in your body . . .

—and that light reaching out of you to touch every aspect of your life, shedding light on your relationships . . . sending light to your friends . . . putting light on any difficult area in your life . . .
—illuminating your whole being, all that is you . . .

—Allow yourself now to experience illumination . . . .

—And taking as long as you need now, slowly begin to find a closure for your meditation, continuing to be aware of the light within you.

1. Swami Abhedananda, *Ramakrishna Kathamrita and Ramakrishna* (Calcutta: Ramakrishna Vedanta Math, 1967), 126.

2. Anthony de Mello, S.J., *Sadhana: A Way to God* (St. Louis: Institute of Jesuit Sources, 1979), 112.

3. Christopher Hills, "Is Kundalini Real?" in *Kundalini: Evolution and Enlightenment,* ed. John White (New York: Paragon House, 1990), 107.

4. Swami Rama, Rudolph Ballentine, M.D., and Swami Ajaya, Ph.D., *Yoga and Psychotherapy: The Evolution of Consciousness* (Honesdale, Pa.: Himalayan International Institute of Yoga Science and Philosophy, 1976), 281–282.

To bring this book to a close I would like briefly to review the practices that you can follow for specific brainwave training. Remember that the intention of these meditations is the mastery of the different states of mind that are reflected by different brainwaves.

The first step toward mastering your brainwaves is learning to relax. Relaxation helps you master all of the other brainwave states. The optimum brainwave pattern, the awakened mind, is a combination of beta, alpha, theta, and delta brainwaves. Rather than going straight into that pattern, it is often easier to first develop the meditation brainwave pattern, which is made up of alpha, theta, and usually delta. To reduce your beta waves there are a number of exercises you can do, including relaxing your tongue, slowing your breathing, and learning to concentrate.

The next step is developing alpha waves. You can do this through any kind of imagery practice, the more vivid the images the better. Remember to involve all of your senses, not just the visual. This sensualization practice will help stabilize the bridge to your subconscious, ultimately improving your memory of your experiences in theta.

You can develop theta by practicing any meditation that takes you down into the depths or up into the heights of yourself. Images such as going *down* stairs, *through* doorways, *under* arches, *along* tunnels, *into* rooms, *up* mountains, *over* bridges, and *around* corners will help take you deeper into theta states. The more such changes you can make in your imagery, within reason, the better.

You can then put the beta waves back into your brainwave pattern by adding a purpose or intention to your meditation. For example, you may go in the boat, down the river, to the island, up the hill, behind the boulder, into the cave, down the tunnel, through the archway, into the octagonal room *where you will find the answer that you are seeking*—or *where the healing that you need will occur*. (I am exaggerating the theta development process so that you get the idea. You don't always need to make it so complex, especially when you are familiar with what the theta state feels like inside yourself.)

You can set the scene for these working meditations by knowing in advance exactly how you want to use your beta waves and what you want to work on. Or you can use more generic beta content by thinking about what you are experiencing in the meditation. It helps, however, to have a framework to work from, so that your mind does not wander off into random, unfocused, high-amplitude beta.

If you don't want beta, simply leave the mental processing out. Arrive at your destination and let yourself meditate.

Developing the radar and sensitivity of your delta waves can be practiced by intentionally reaching out to others with your empathy, interest, insight, and awareness.

formula for creating your own brainwave-training meditation

1. Identify the purpose or goal of your meditation, if there is one.
2. Start with relaxing your body.
3. Clear your mind and reduce beta waves.
4. Develop alpha waves through imagery, perhaps sensualizing an environment in which to experience your meditation.
5. Develop theta waves by going deeper in your internal journey, using key images and sensations, as described above.
6. Add beta to your meditation by processing the content in some way (for working meditations only).
7. Crystallize your experiences in meditation by finding keys to help retain your memory of the content.

8. Close your meditation and ground it by speaking, writing, or even just thinking about your experience.

Following this structure, you can create an infinite number of meditations for any purpose, using imagery and experiences that work well for you.

Before you close this book, take a moment to consider what you have discovered. Close your eyes and reflect on what was the most important information for you, what you can take away with you from reading this book, and how you can integrate this knowledge into your daily existence. Imagine how you can benefit from the practices, ideas, and meditations that were the most significant for you, and sensualize making use of them regularly in your life.

Remember the image or symbol that you found in the first meditation, at the end of Chapter One, which represented where you were in your life at that point. You may wish, once again, to allow yourself to find a symbol for where you are in your life right now. Notice the similarities to that first image, and the differences.

Take a minute to look into the future. Imagine the changes that can take place in your life as you continue to work at developing your high-performance mind.

APPENDIX

•

Kundalini

Developing a high-performance mind requires personal commitment. People who wish to achieve this goal have learned to integrate tools for brainwave mastery into their lives, creating a meditation routine that is both enjoyable and tailored for their individual needs.

Once someone has found his or her rhythm with meditation and developed a regular practice, long-term effects may begin to take place. These effects vary greatly and depend upon the individual. Some results come quickly; others may take time to mature.

Among these long-term effects you may find greater skill in handling unexpected situations; more mental flexibility and increased creativity; greater intuition and insight; fewer health problems, or a better ability to deal with the health problems that do occur; increased emotional stability; less overall stress; and a heightened spiritual awareness.

If you meditate regularly, you will almost certainly feel changes. These changes can be as subtle as a gentle sense of relaxation, or they might manifest in feelings of serenity or even dramatically as periods of ecstasy and bliss. You may also experience mood swings or feelings of confusion or sadness as you undertake your subconscious housecleaning. With practice, you will learn how to use even the downtimes beneficially.

Although most people who meditate experience beneficial states, you may go through low periods as well as high. You may feel fatigue as well as increased energy. Your dreams may change, as well as the quality of your sleep. The timing of the effects of meditation is different for everyone. If you hit a low period, bear in mind that all of the changes are ultimately leading toward the positive—toward the development of your mind and the evolution of your being. If there are some dark rooms inside to clean out, or you discover personal changes that you need to make, welcome these as opportunities to further your journey.

A very small percentage of people, however, experience much stronger effects from intense long-term meditation. These effects may be heightened by their own journey of internal searching and evolution. Although these people are going through a completely natural process, they may feel the effects of their awakening in stronger than usual, even dramatic ways. Because meditation is a powerful tool for transformation, I feel a moral and ethical obligation to discuss *all* possibilities with you, in case you should find yourself experiencing any of the symptoms below.

While I was experiencing a personal awakening, I had not been forewarned of the possibilities of the physical, emotional, and energetic nature of the process. It would have saved me months of worry to have had this explanation. Therefore I offer this Appendix for those readers who are interested in, or are experiencing the effects of, kundalini.

KUNDALINI

Very occasionally, long-term meditation can lead to emotional swings and dramatic physical sensations as the body begins to reorganize its energies. Whenever you work on yourself, in whatever discipline or modality you choose, you will find changes beginning to take place within yourself if the work is effective. One of the changes that can occur through long-term meditation is an activation of what is known as kundalini.

Every spiritual tradition has some form of "rebirth" process that signifies the individual's spiritual awakening. This is often characterized by the unconscious becoming conscious. There are very few spiritual traditions that describe this physiologically as well as psychospiritually. In the yoga philosophy, this rebirth is called the awakening of kundalini.

The kundalini is symbolized by a serpent. *Kunda* means literally "bowl" or "basin." Before its awakening the serpent is coiled in the pelvic bowl. The kundalini serpent is the storehouse and spiritual link to a wealth of energy, powers, abilities, and awareness that, when released, are said to mark the first steps toward enlightenment. The activation of kundalini is the *release* of this latent energy into the individual's physical, emotional, mental, and spiritual system and the beginning of a wide variety of phenomena. When the process is completed there is psychological maturity, emotional balance, personal strength, and spiritual awakening. Because it is a rebirth, however, in the initial stages of the process the individual experiences all of the confusion, helplessness, fear, and unpredictability of a child. The process, then, is one of *allowing* this energy to develop and mature.

You do not create *the kundalini energy,* just as you do not create your own life force as a child, but you can tap into it, learn to master it, and encourage it on its journey of development. You can also hinder it, block it, distort it and misuse it.

THE PROCESS OF EVOLUTION

I firmly believe that the development of the awakened mind is an evolutionary process as well as an individual process. At some future time, some evolved form of the human race may look back on the "Dark Ages" of the first twenty or so centuries, before humankind became awakened. They will look back at a time when people's minds were still closed to one another, when people's feelings and experiences were unreachable, hidden deep within them, a time when we truly did not know the mysteries of the spirit and creation.

I also believe that the awakening of kundalini energy is an evolutionary process that is becoming more and more common with each passing decade. Lee Sannella, M.D., provided our Western minds with a framework through which to view this awakening as a physiological process in his book *Kundalini, Psychosis or Transcendence,* later published as *The Kundalini Experience*. This book became a great help to me in the late 1970s in London, when I was undergoing the most violent of my own kundalini symptoms.

The awakening of kundalini may not necessarily be tumultuous. It may be so gentle and prolonged as to be almost unrecognizable. Awakening takes place on many different levels. It is by no means necessary or even common to have a kundalini awakening accompanying the development of a high-performance mind. But it is possible, just as it is possible to have a kundalini awakening without *ever* practicing any form of meditation or personal and spiritual growth.

The process of kundalini can feel like both a blessing and a curse. The blessings are increased awareness and perceptions; access to potent energies that can be applied in many different directions, such as creativity, healing, and the service of others; personal stamina, health, and well-being; expanded spiritual understanding; and emotional and mental clarity and balance. It may also feel like a burden, as abrupt emotional swings catapult one into the dark night of the soul, or unexplained physical sensations and pains lead doctors to use labels such as hypochondriac, mentally unstable, or hapless victim of some fatal disease. Long sleepless nights interrupted by periods of violent shaking can evoke serious questioning of both one's spiritual path and one's sanity. These are the very rare extremes.

This kind of awakening can lead to what Stanislav and Christina Grof have called a "spiritual emergence" or "spiritual emergency." My experience ranged the full extent of the pendulum swing between bliss, ecstasy, joy, and union with the cosmos to pain, despair, confusion, and fear. It wasn't until after I had spent months worrying about my condition that I thought to tell my teacher, Max Cade, what was going on.

Max was very compassionate and somewhat regretful that he had not informed me of this process earlier and thereby prevented a year of anxiety, fear, and uncertainty. He gave me Dr. Sannella's book to read and told me everything would be all right in time. Not surprisingly, the symptoms lessened almost immediately. *Acceptance* of the phenomenon and some theoretical understanding of it allowed me to let go and stop trying to control it. When I let go, the energy was able to move much more freely through my body, the blocks were less painful, and the symptoms less disturbing. I believe that all health care providers, especially those focused in the area of the nervous system or the energetic system, should be trained to recognize the symptoms enough to at least be aware of the possibility of kundalini awakening and to refer the patient to someone skilled in that field.

SYMPTOMS OF KUNDALINI AWAKENING

So let us now look at the scope of symptoms that *can* occur. Please remember while reading this that these do not *all* occur and they may occur at different times, spaced very far apart. Some people only lightly experience a few of the symptoms yet are still undergoing a kundalini awakening. Some experience one type, then get a break before experiencing a completely different set. Others may go through the whole range quickly or slowly.

Physical Sensations

The movement of the energy can cause the feeling of crawling or itching on the *inside* of the skin. A sensation of tingling, pins and needles, or numbness, sometimes accompanied by a feeling of inner vibration, bubbling, or quivering, can appear anywhere in the body but may be especially prevalent in the limbs, often starting in the feet and legs.

Because symptoms often begin in the root chakra, in compressed nerves at the base of the spine, the initial sensations may begin in the lower extremities, frequently the toes. Numbness of the anal area is also possible, as the first chakra—in yogic traditions located in the perineum—undergoes its "opening." In some Tibetan traditions, however, the first chakra is believed to be in the soles of the feet, which would also explain the tingling in the toes. As the energy moves along its path, these sensations can move throughout the body, either in the orderly fashion of the classic kundalini awakening or in a seemingly random and inconsistent way. Pain is also a common symptom, ranging from muscular aches and pains to painful joints to severe headaches.

Body Movement

This symptom can be as benign as simple rhythmical movement during meditation or as extreme (and rare) as spontaneously standing on one's head, without conscious choice or intention. Spontaneous yoga positions are common as the energy pushes the body into specific positions in its attempt to open up the channels of free flow. Yoga was not developed by some ancient sage sitting down and *thinking* that these would be beneficial postures to practice. Yoga was developed by the emergence of the energy itself, placing the body in those postures. The master, as he awakened, practiced those positions intuitively and spontaneously, and the student observed him and copied, wanting the same awakening for himself.

A number of times I woke up in the middle of the night in the Yoga position of the bridge. The first few times this happened, it was quite a surprise. When I knew what was happening and did not try to resist the energy's movement, it began to happen more frequently. I got to where I welcomed the pleasurable sensations and release this position gave me.

Convulsive or spasmodic movements can also take over the whole body. If this happens, try to be aware of *the wave* of energy as it is pulsing through you and allow yourself to ride that wave. Then what originally might feel like a convulsion can become a smooth rhythmical body movement. Reichian bodywork has been tapping this energy for decades. The bottom line here is, don't try to stop the movement. Instead, relax around it until it becomes organized, rhythmical, and smooth.

Finally, spontaneous stretching can also occur, most frequently in times of rest and repose. You may experience a sense of elongation, especially of the spine and back. If possible, let yourself stretch without straining or tensing.

Visual Symptoms

These range from brief flashes of light inside the head to full-blown, lengthy spiritual visions to frightening hallucinations. Many times I was awakened by what I can only describe as a light bulb going off in my head. One might also experience the "seed of light," a tiny point of light that appears in the third-eye area, between the eyebrows. Sometimes referred to as the blue pearl, this light will seem to appear between you and anything that you are looking at. While the seed of light may first be seen when your eyes are closed in meditation, or through practicing the tantric candle meditation "Trataka," it can continue to appear spontaneously while you are in an ordinary waking state with your eyes open.

You may experience temporary visual problems during this process, especially if you already have certain weaknesses in that area. Though rare, spontaneous blindness has been known to occur, a condition which should reverse itself rather quickly. My own path through vision problems would take a whole chapter to describe. Although my condition was eventually diagnosed as ocular histoplasmosis, I know that the process of its manifestation was intricately related to my energetic awakening.

Illumination

I give this state its own category because of its uniqueness. Some rare individuals have been known to glow visibly in the dark as a result of this awakening. Perhaps this is the source of the halo in religious art. The halo may not simply be an artist's rendition of a representation of spirituality, but a visible illumination caused by light emanating from the crown chakra and the circle of chakras around the top of the head.

Auditory Symptoms

Hearing can become distorted. A wide range of inner sounds can occur, including ringing, hissing, crackling, buzzing, whistling, whooshing, roaring, knocking, ocean sounds, and musical notes. One might occasionally hear voices, sometimes identifiable as your own voice, or at other times more like an auditory hallucination. These symptoms are especially noticeable during meditation or in the quiet of the night, when they can escalate to a loud roar.

Temperature Changes

The sensation of extreme heat might manifest itself in the body in two ways. The energy moving through the body can cause a sensation not unlike a hot flash. It is also possible for the extremities of the body, especially the hands, to become excessively hot to the touch. While very rare, cases have been recorded where a burn mark appeared on the furniture an individual had touched.

Changes in Breathing Patterns
This most frequently occurs during meditation. The breathing patterns can range from short, sharp, and shallow to long, extended, slow breaths almost resembling the cessation of breathing.

Emotional Swings
During an energetic awakening, there is absolutely no emotion that is off limits or out of bounds. You might experience sensations of bliss, harmony, peace, and universal joy for increasingly longer periods of time, or you can just as easily swing sharply into feelings of despair, low self-esteem, fear, or even insanity. Self-doubt is very common and takes many forms. You may doubt your own reality. You may doubt that these things are really happening to you. You may wonder, "Why me? Why now?" You may feel as if you are making it all up. You may also have occasional feelings of grandeur, arrogance, and vanity that might earmark energy blockage. A feeling of isolation stemming from the assumption that you are different from everyone else is also common and can be alleviated by finding a support network of some sort, or at least one other individual who has had similar experiences.

Perceptual Changes
A sensation especially common during meditation is the sense of growing to great size, expanding beyond your body, or having an out-of-body experience. In a milder form this state can be felt as extreme detachment or disassociation.

Changes in Sexual Energy
Because of the nature of the human energy system, kundalini energy is intricately tied up with sexual energy. They are not, however, the same thing. Because they are both types of energy, they stem from the same root, but kundalini is much greater than sexual energy. It

would be more accurate to consider sexual energy as one aspect of kundalini. Sexual intercourse or orgasm can temporarily stimulate and increase the activity of the kundalini energy. After an orgasm, you may feel increased symptoms in any of the categories discussed above, especially body sensations and body movement.

The awakening of the kundalini may also stimulate an increased sexual desire. The arousal of the energy itself can feel like an orgasm. When it is complete and moves up above the lower sexual chakra, it is experienced as a total-body orgasm, much fuller and more ecstatic than a genital orgasm. If you are meditating on the awakening of the kundalini and have started the process, you may find that you come to orgasm during a meditation without movement, stimulation, or imagery of any kind. One female colleague reported her embarrassment at climaxing unexpectedly during her weekly meditation class.

THE HOSE IN WINTER

The process of kundalini awakening is one of purification and the de-stressing of the entire psychophysical system. In order for this purification to occur the system has to be cleared of blocks in the energy channels so that the energy can flow smoothly, evenly, consistently, and fully throughout. Among the many causes of these blocks are physical disturbances, emotional trauma, illness, diet and nutrition, and mental and psychological belief systems. These can be internal messages carried throughout life, such as "It's not O.K. to feel pleasure" and "Higher spiritual awareness does not exist."

Think of your body as a garden hose left outside over the winter. Twigs and dirt have made their way inside. Perhaps there are even small stones or pebbles to block the water's flow. The hose has been kinked in several places, as well as stepped on and crushed.

Because you didn't need to water anything in the winter, you haven't been concerned about the hose. If you turned on the water, only a trickle would come out unimpeded, and perhaps that would have been all you needed. What happens when you turn the water up full force? The hose flops around until it finds a stable and open position to accommodate the unaccustomed strong flow of water.

The blocks are blasted out by the force of the flow, and eventually the stream of water is strong and the hose is fully functioning.

The human energy system, while decidedly more complex, can behave in a similar fashion. When a strong surge of kundalini energy flows through obstructed channels, the body tends to flop around just as the hose does, hence the origin of the *kriyas*—yogic "actions," or movements.

The classic kundalini awakening initially involves the rising of energy from the root chakra in the perineum. As I mentioned before, the earliest symptoms are often experienced in the feet, the legs, and the toes—the big toes in particular.

As the process continues, the energy rises up through the chakras, clearing any obstructions it finds in its path. As it hits a chakra, it tends to spread out through all of the psychophysical systems related to that chakra. When the energy encounters a block or resistance and then overcomes it and cleanses the system governed by that chakra, that chakra is considered to be opened.

Traditionally the energy progresses in sequence, from the root chakra up to the crown chakra in the top of the head. Then it spreads down across the face and the chest and comes to rest in the navel area. In practice this orderly progression rarely happens. The movement of the energy jumps around from location to location. You may indeed feel it first in the feet and legs, but then feel it next in the throat, before it goes back down to the heart.

The energy may seek out the largest blocks first and work in layers. The cycle of awakening may happen again and again as layers of resistance are overcome and the energy system is more and more refined. The time this process takes is also widely variable. The major portion may be completed in a few years, or it may continue slowly for decades. It is also possible to feel the process is complete, only to have it recur years later around a new block that has developed or as a further experience of fine-tuning.

Looking for an end to the experience does no good whatsoever. It is a *process*, not a *goal*.

SPONTANEOUS *SHAKTI-PAT*

Shakti-pat is the direct transmission of energy from one who has the energy flowing freely to one who has not yet realized the awakening of kundalini. The most common occurrence of *shakti-pat* takes place during the passage of energy from spiritual teachers to their students. *Shakti-pat* can occur through touch, through look, through presence, or even long-distance through a form of thought transmission. Muktananda, founder of siddha yoga, was one of the most well known purveyors of *shakti-pat*. Some of his students had the experience of being forever changed by his mere presence or gentle touch. All true spiritual teachers practice some form of transmission to their students—such is the nature of teaching. Tantric yoga masters such as Swami Satyananda Saraswati have been sought after because of their gifts at assisting meditators with the energy-awakening process. The medium of these spiritual gifts need not be in the form of kundalini energy. Many spiritual teachers teach by principles such as devotion, right action, and being fully present in the moment, and leave the kundalini to take care of itself.

Shakti-pat does not occur only through the medium of a spiritual teacher. Anyone who has the energy awakened in his or her own system can, under certain circumstances, transmit that energy to others and assist in their awakening—even at times without knowing it. Kundalini energy "knows" kundalini energy. If you are in the process of a kundalini awakening and spend a few minutes with another person in the same condition, you will most likely recognize each other.

COMMON PLACES TO GET STUCK

In most instances, the energy has a mind of its own and the process is one that unfolds beyond your ability to really control it. This process *can* be helped or hindered, however, by the actions and attitudes of the individual undergoing it. There are a few areas where people get stuck by encouraging the energy to stay in its current location, or block, rather than following its natural progression. Some of these are outlined below.

The Pitfalls of Pleasure

One of the most common problems I see in clients who have come to me for kundalini counseling involves a phenomenon that is often misunderstood. These people are usually well in the throes of the kundalini awakening. They profess to understand it (somewhat) and to accept it. They are often meditators and frequently have a fairly stable awakened mind brainwave pattern when they meditate.

When they meditate, however, their bodies go into uncontrollable movements—stretching, bending, shaking, trembling, and large wavelike, orgasmic motions. They have a love-hate relationship with these phenomena. On the one hand they are embarrassed by the experiences, especially the outer manifestations of the energy movements. They feel isolated and misunderstood. They cannot meditate in a group because their contortions are disturbing to the other meditators. And nothing they do seems to change this phenomenon or prevent it from occurring. On the other hand, when these sensations happen, they experience intense physical pleasure. They *know* that it is a kundalini based experience, so they feel secure in the knowledge of their spiritual evolution.

These people are caught in a catch-22 that may inhibit their progress for a long time unless they take steps to break the cycle. The energy moving through them is too great for the channels it is moving through. Or, alternately, the channels are too small to accommodate the amount of energy. This is typical for all kundalini awakening, and there are only two things you can do to alleviate this problem—reduce the amount of energy or increase the size of the channels. These people are doing neither.

As they feel the pleasure of the friction of the excess energy moving through their inadequate channels, their bodies move accordingly. This is their feedback that they are really having a kundalini experience, not just imaging it or wishing it would happen. The mistake they make, however, is in thinking that these body movements are something to be sought after, something to be valued and maintained at all costs. So when they feel the rush of energy coming, they automatically squeeze the energy channels even tighter around it, thereby increasing the pleasure of the friction, perpetuating the body movements, and reinforcing the certainty that they are indeed having a kundalini experience. Their feedback loop is complete. The individual continually inhibits the energy from progressing on its natural journey by creating intentional blocks, because this provides proof that the kundalini process is in action, as well as giving them momentary pleasure.

This cycle can happen repeatedly. It will end naturally when the individual gets tired of the awkward physical movements and when the pleasure and glamour of the phenomenon diminishes. But it can also go on for a very long time.

The solution here is education and acceptance. The individual needs to learn that the movement and pleasure are not the end goals —only a passing phase of the whole kundalini process. He or she needs to accept that the kundalini experience is really happening, and let go of the need for dramatic, external proof. Only then can he or she allow the energy to take its natural course. The physical manifestations may not stop immediately, but they will diminish and cease over time.

The goal in this part of the process is to be able to sit quietly and channel an increasingly larger amount of energy without being imbalanced by it. Only in this way will the amount of energy that can be handled by the system increase.

It is important to understand at this point that just as the unintentional body movements are not good and something to be sought after, neither are they bad and something to be studiously avoided. They are just one of the things that happen in a kundalini awakening, and they are to be accepted, experienced, and then progressed beyond. You can even allow them to be exciting if you wish— something new, interesting, and pleasurable—as long as you don't get stuck in maintaining them.

Ego
Another major stumbling block for some people is difficulty in dealing with the fact that this is happening to *them*. They take it personally and may begin to see themselves as better than others or more

evolved. As soon as this kind of personal ownership sets in, the individual is headed for trouble. Kundalini does not progress in the face of this kind of possessiveness. The ego itself becomes a block that the kundalini energy must work through in order to move forward. The more the ego tries to hold on to its inflation, the harder the kundalini has to work to do its job of purification. The unfortunate result of this conflict may be an individual who proudly proclaims, "My ego is smaller than yours!"—a classic contradiction in terms.

Whenever anyone deals with an ego problem, whether kundalini based or not, self-esteem is an issue. Self-esteem problems are one of the blocks that the kundalini energy must work through on its journey of awakening.

THE SELF-ESTEEM PENDULUM

I see self-esteem problems as a pendulum. At one end of the pendulum is an overinflated ego, and at the other end is low self-esteem. Usually, when you have one problem, the other is lurking nearby. All it takes is a simple swing of the pendulum to manifest the extreme opposite.

Many people think that the solution to this swing between ego extremes is to find a nice safe balance in the middle. *Absolutely not.* The only real solution is to *get off the pendulum altogether!* If you are trying to find a balance, you are still battling with the polarities of both extremes, even if you find and maintain your balance for an extended period of time. The potential is always there for the scales to tip and the pendulum to swing to one end or the other, or both in quick succession, which is often the case.

So how does an individual get off the ego pendulum? Detachment and letting go are the main answers—easy words to say, but difficult concepts to comprehend. The easiest and most direct way to practice detachment is to simply accept that there is a power greater than yourself that is ultimately in charge of everything. *YOU are not in control.* You did not cause the kundalini awakening, and you do not own the energy. And you are not special because it is moving in you. It is simply what is happening.

This book is not meant to be religious, but when we talk about awakening, we cannot avoid spirituality. It doesn't matter what you call that divine power; it only matters that you do not see yourself as the highest authority. If you see yourself as the final authority in everything, you are headed for a fall, and the block in your kundalini awakening could be long and painful. The more faith you have in some divine order in the universe—whether your higher power is God with a capital G, nature, source, Buddha, Jesus, love, or light— the easier it will be for you to let go of your own ego difficulties.

Fear

A third major stumbling block in the kundalini process is fear. This is easy to understand because many of the experiences that occur involve working through and releasing the deepest, darkest blocks and resistances in your body-mind. Kundalini is about releasing stress and the purification and cleansing of a lifetime of personal demons. If your demons are dark (and whose aren't?), you are liable to experience fear.

Faith in a power greater than yourself is a source of great strength here. If you allow yourself to be overwhelmed by the fear, you will perpetuate and even increase it. If you understand that it is only a passing phase of the process as a whole and let it go, just as you did the ego, then you can move on and the process can continue in a manner that is appropriate for you.

HELP!

An important question you may be asking now is, "What can I do if this process starts happening to me? How can I get the help or guidance that I need?" There are many advertised kundalini classes and practitioners who purport to be aware of the process. Search them out and ask for their help. You might, however, run into some problems here. However empathetic and well intentioned these individuals are, unless they have experienced the energy transformation personally and gone through (or at least are in the process of) their own kundalini awakening, their help may not be what you need.

Seek out someone who understands the phenomena from the inside, from his or her own experience.

Kundalini knows kundalini. Because it is an impersonal and universal energy, it recognizes itself in another human being. Within a matter of hours, if not minutes, of being with someone who has undergone the process, you will know that he or she understands what you are experiencing. You share a common language. I do not recommend studying kundalini with teachers who are teaching it because they are trying to activate it within themselves. Let your guides be at least one step ahead of you, and they will help you down your path with greater ease, clarity, and wisdom.

Christina Grof began the Spiritual Emergence Network (SEN), which is an information and referral service for people undergoing all kinds of spiritual emergencies including kundalini. Call (408) 426-0902 for a reference to a practitioner near you.

SELF-HELP

Educating yourself about what you are experiencing as well as *finding another person on the same path* will provide support through the more difficult periods. Reading books about kundalini to familiarize yourself with the process and learning what others have done to help themselves is a great source of knowledge *as long as you take what is helpful and leave the rest.* Everyone's experience is different. While the real key is to come to terms with *your own experience,* learning about the experiences of others can help greatly.

How you physically care for yourself also has a great impact on your kundalini experience.

Diet
The food that you eat affects the energy systems in your body dramatically. Many people undergoing a kundalini awakening will find that their diet changes naturally without any conscious decision to change it. My own experience with food was extreme, but only for a matter of months.

Before my energetic awakening began, I was an ordinary meat eater, taking many of my meals in restaurants because of my busy lifestyle. When I was going through the most dramatic initial phases of the energetic changes, I was unable to eat meat. I say unable because this was not a mental or emotional response to meat; it was a physical response. I actually could not swallow meat without gagging on it and feeling nauseated.

I became a gourmet vegetarian, cooking nut roasts and elaborate cheese dishes. Then dairy products became difficult. Finally the nuts and grains became too heavy for me, and I found myself able to consume only vegetables. My diet turned towards raw foods. Finally, at the peak of my dietary disturbance, I spent about a month feeling comfortable eating practically only lettuce. This time was also interspersed with periodic fasting. I felt best during the fasting because my energy flowed the most easily and clearly. Gradually, as that phase of the kundalini symptoms died down, heavier foods came back into my diet. But they came back in with a difference. While I am now capable of eating most foods, I am comfortable with only very *clean* food. I find eating in many restaurants not nearly as pleasant as it was before, and my system cannot tolerate junk food. Large salads, seeds, nuts, grains, fruits, vegetables, and pastas make up the bulk of my meals.

I have written about my personal diet as a way of explaining that what is best to eat during your kundalini experience is not necessarily a conscious decision. What is required is a willingness to follow what your body wants and to listen carefully to your own needs.

In general, if you want to slow down the kundalini process, eat heavier foods. If you want to speed it up, eat lighter foods. I suggest staying away from toxic or fast foods at all times during this process. Let your body be your guide. Maybe a good greasy hamburger will be just what you need one day to ground you!

CASE STUDY—T.K.

T.K., a man in his early forties, had been meditating for many years. He had been undergoing a kundalini awakening for over a year when he came to visit me from the East Coast for a brainwave profile and help in making the energy more a part of his life. He frequently experienced states of bliss and spiritual illumination during meditation. Able to enter these states instantly, he found little difference between the state of meditation and the rest of his life.

Figures 1A and 1B show the most common brainwave patterns from his waking profile. T.K. produced a clear, stable awakened mind pattern that would dissolve within a minute and then return the next minute. These cycles continued consistently for the duration of the waking portion of the brainwave profile.

Figures 2A, 2B, and 2C depict the three fluctuating patterns he presented during the meditating portion of the profile. He moved between a solid meditation brainwave pattern (Figure 2A), an awakened mind pattern (Figure 2B), and a transitional pattern of too much beta, strong alpha, almost no theta, and a little delta (Figure 2C).

T.K. knew without a doubt that his kundalini was active and his mind was awakening, although he was grateful for the Mind Mirror EEG confirmation. His main problems were with the *kriyas* that his body went into whenever he meditated and occasionally when he was not in meditation. His body gyrated strongly, rocked back and forth, and shook uncontrollably. His head rolled back, and strange sounds came from his throat. While he found these experiences pleasurable—sometimes exceedingly so—he was unable to meditate in his "temple" with his spiritual group because of the disturbance he caused the other members. He found himself increasingly isolated and unable to explain his actions and his feelings to his friends and associates.

T.K. found immediate relief in my acceptance and understanding of his problem. He also felt validated by the EEG assessment that he was on the right track and was not a fraud. The main work that we needed to do was in the area of relaxing and broadening the energy channels and *allowing* the flow of energy to come without the violent movements. He discovered he had been *clenching* the energy to embrace it, hold it, and feel it; and the more he clenched, the more violent his movements became.

Once he realized that having the *kriyas* was not a necessary or even desired component of his energetic process, he was able to begin to release his stranglehold on the emerging energy and let it flow through his body without as much physical effect. Seeing his awak-

CASE STUDY T.K.

NORMAL WAKING STATE

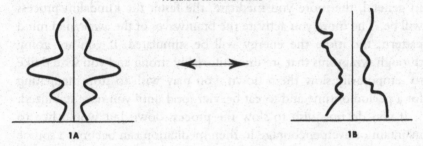

1A   1B

T.K. MEDITATING

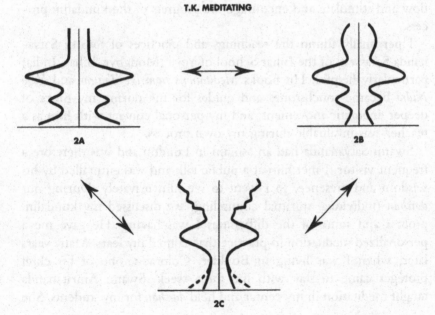

2A   2B

2C

ened mind brainwave pattern on the EEG also allowed him not to need the validation provided by the violent movements anymore. He was able to continue his process with more ease and certainty.

MEDITATION

Meditation is one way of helping to regulate the kundalini energy. In general, the more you meditate, the faster the kundalini process will be. The more you activate the brainwaves of the awakened mind pattern, the more the energy will be stimulated. If you are going through symptoms that are uncomfortably strong and you would like to temporarily slow them down, you may wish to stop meditating for a period of time and to eat heavier food until you have stabilized.

If you do not wish to slow the process down but would like to maintain or even encourage it, then meditation can become a source of great strength and support for you. Any form of meditation will support the kundalini process because it alters the brainwave patterns to those that can help stimulate awakening. There are specific forms of meditation that you can practice to stimulate kundalini. Various tantric practices work on the chakras to help increase the energy flow and stimulate and encourage the progress of the kundalini process.

I personally found the teachings and practices of Swami Satyananda Saraswati of the Bihar School of Yoga (Monghyr, Bihar, India) particularly helpful. His books *Meditations from the Tantras* and *Yoga Nidra* became touchstones and guides for me during my times of deeper energetic movement, and my personal contact with him as a teacher was invaluable during my own process.

Swami Satyananda had an ashram in London and was therefore a frequent visitor. I met him at a public talk and was enthralled by his wisdom and presence, so I went to see him privately. During our *darshan* (individual spiritual counseling) we discussed my kundalini process and some of the difficulties I was having. He gave me a personalized meditation to practice that calmed my fears. Many years later, when I was living in Boulder, Colorado, one of his chief protégés came to stay with me for a week. Swami Amritananda taught meditation in my center and held *darshan* for my students. She and I spent time together, during which she blessed my soon to be born child and gave me a great deal of encouragement.

I began using some of Swami Satyananda's meditations in my teaching in London. I introduced his writings to Max Cade, who was so impressed that he also integrated them into his own work.

Although a traditional tantric teacher, Satyananda also wrote about the brain and the effects of practicing yoga on the brainwaves, so it fit in well with what Max and I were doing. While his EEG work was rudimentary and still somewhat speculative on the meaning of the different brainwave rhythms, Satyananda nevertheless completed the marriage between consciousness, meditation, and brainwaves in a way that few other people were doing at the time. In 1974, he wrote:

How can one consciously change the brainwave emissions? At first this seems impossible, for the wave form is a function of the mind itself. It might seem surprising that all that is required is practice and perseverance, but this is exactly the same thing that we do with everything in life . . . [Biofeedback] will complement meditation and will bring meaningful meditation experiences into the range of all people. Meditation coupled with biofeedback could do much to raise the level of the mind, or the level of consciousness in the world.[1]

It's no wonder that he was my earliest Indian teacher.

1. Swami Satyananda Saraswati, *Meditations from the Tantras* (Monghyr, India: Bihar School of Yoga, 1974), 23.

More Resources from
Anna Wise

Anna Wise has created several audiotaped meditations, including a set of audio-cassettes based on the meditations in this book. These work with both the state and the content of consciousness for high-performance mind training. She has also designed music tapes for brainwave development that contain complex sequences of frequency tones to encourage the activation of a high-performance mind.

Information about seminars, workshops, and practitioner training programs is available on request.

To see a schedule of upcoming events, or for information about obtaining CDs or other products, including the Mind Mirror III EEG, please visit our website at www.annawise.com or contact us at info@annawise.com

CDs BY ANNA WISE

High Performance Mind
Four-CD set of 10 guided meditations from this book that work with both the state and the content of consciousness for high-performance mind training. All meditations are backed by specially designed frequencies and music that encourage the brainwaves desired.

Meditation Frequencies and Music
Music CD for meditation and brainwave development that contains complex sequences of binaural beat frequency tones to encourage the activation of specific meditation brainwave frequencies. Music by Kit Walker. Frequencies by Anna Wise.

Awakened Mind Frequencies and Music
Music CD, with specific binaural beat frequencies interwoven that encourage an awakened mind brainwave pattern. Sequence is five minutes long and may be used as a quick "tune-up" before a creative project or important meeting, or can be repeated as background music to accompany situations where an awakened mind is needed. Music by Kit Walker. Frequencies by Anna Wise.

Sound Body, Sound Mind: Music for Healing, **with Andrew Weil, MD**
Two-CD set with booklet. CD 1—A Symphony of Brainwaves, psychoacoustic music combined with beat frequencies guiding you to a place where healing can occur. CD 2—Narratives by Dr. Andrew Weil on healing and sound; Anna Wise on frequency development; and Joshua Leeds on the making of Sound Body, Sound Mind. Produced by Joshua Leeds. Frequencies designed by Anna Wise.

LATEST BOOK BY ANNA WISE

Awakening the Mind: A Guide to Mastering the Power of Your Brain Waves
Jeremy P. Tarcher/Penguin, 2002. Developing the brainwaves for self-exploration and healing that lead to the intricate synthesis of the qualities of mastery. The fine points of awakening your mind, including kundalini and the expansion and contraction of consciousness. Dr. Andrew Weil says of this book, "A first-rate guidebook for the development of consciousness, by one of the foremost pioneers of brainwave biofeedback. *Awakening the Mind* is filled with practical secrets that anyone can use to make fuller use of the mind and become more skilled at living."

SEMINARS, WORKSHOPS, AND TRAINING PROGRAMS

Esalen Institute Five-Day Seminars
The High-Performance Mind: Awakened Mind Brainwave Training
Awakening the Mind: Mastering the Power of Your Brainwaves
EEG and Spirituality
Advanced Awakened Mind: Biofeedback, Meditation, and Consciousness
  Training (weekend)

ADVANCED WORKSHOPS

Brainwave Training and Healing
The Expansion and Contraction of Consciousness
The Awakened Mind and Kundalini

PRACTITIONER TRAINING PROGRAM

An in-depth series of four five-day seminars and two practicums designed to teach you to help people awaken their minds. Brainwave pattern recognition, interpretation, and training, profiles, signature patterns, individualized protocols, individual sessions and the content of consciousness, Awakened Mind group training and the Wise Protocol, writing and leading meditations, group and personal "energetics," spiritual crisis, and spiritual development.

CONTACT INFORMATION—WEBSITE

www.annawise.com

BIBLIOGRAPHY

Beattie, Melody. *Beyond Codependency: and Getting Better All the Time*. San Francisco: Harper and Row, 1989.

———. *Codependent No More: How to Stop Controlling Others and Start Caring for Yourself*. San Francisco: Harper and Row, 1987.

Blundell, Geoffrey G. *The Meaning of EEG: A Study in Depth of Brain Wave-patterns and Their Significance*. London: The Publications Division of Audio Ltd.

Blundell, Geoffrey G., and C. Maxwell Cade. *Self-awareness and E.S.R.: An Extended Study into the Measurement of Skin Resistance as a Guide to Self-awareness and Well-being*. London: The Publications Division of Audio Ltd.

Cade, C. Maxwell, and Nona Coxhead. *The Awakened Mind: Biofeedback and the Development of Higher States of Awareness*. Shaftesbury, Dorset, Great Britain: Element Books, 1989.

de Mello, Anthony S.J. *Sadhana: A Way to God: Christian Exercises in Eastern Form*. St. Louis: The Institute of Jesuit Sources, 1979.

Funderburk, James. *Science Studies Yoga: A Review of Physiological Data*. Himalayan International Institute of Yoga Science and Philosophy of USA, 1977.

Grof, Stanislav and Christina (eds.). *Spiritual Emergency: When Personal Transformation Becomes a Crisis*. Los Angeles: Jeremy P. Tarcher/Perigree, 1989.

Hills, Christopher. "Is Kundalini Real?" *Kundalini: Evolution and Enlightenment*. (John White, ed.) New York: Paragon House, 1990.

Joy, W. Brugh. *Joy's Way: A Map for the Transformational Journey: An Intro-*

duction to the Potentials for Healing with Body Energies. Los Angeles: Jeremy P. Tarcher, 1979.

Motoyama, Hiroshi, with Rande Brown. *Science and the Evolution of Consciousness: Chakras, Ki and Psi.* Massachusetts: Autumn Press, 1978.

Ornstein, Robert, and Richard F. Thompson. *The Amazing Brain.* Boston: Houghton Mifflin, 1984.

Restak, Richard. *The Brain.* New York: Bantam Books, 1984.

Riviere, J. Marques. *Tantrik Yoga:* Hindu and Tibetan. (H.E. Kennedy, trans.) New York: Samuel Weiser, 1971.

Sannella, Lee. *The Kundalini Experience: Psychosis or Transcendence?* Lower Lake, California: Integral Publishing, 1987.

Simonton, O. Carl, Stephanie Matthews-Simonton, and James L. Creighton. *Getting Well Again: A Step-by-step Self-help Guide to Overcoming Cancer for Patients and Their Families.* Toronto/New York/London: Bantam Books, 1978.

Swami Abhedananda. *Ramakrishna Kathamrita and Ramakrishna: Memoirs of Ramakrishna.* Calcutta: Ramakrishna Vedanta Math, 1984.

Swami Rama. *Living with the Himalayan Masters: Spiritual Experiences of Swami Rama.* (Swami Ajaya, ed.) Honesdale, Pennsylvania: The Himalayan International Institute of Yoga Science and Philosophy, 1980.

Swami Rama, Rudolph Ballentine, and Swami Ajaya. *Yoga and Psychotherapy: The Evolution of Consciousness.* Honesdale, Pennsylvania: The Himalayan International Institute of Yoga Science and Philosophy, 1976.

Swami Satyananda Saraswati. *Meditations from the Tantras.* Bihar, India: Bihar School of Yoga, 1983.

Swami Satyananda Saraswati. *Sure Ways to Self Realisation.* Bihar School of Yoga, Bihar, India, and Satyananda Ashram, Australia, 1984.

Tellington-Jones, Linda, and Sybil Taylor. *The Tellington TTouch: A Breakthrough Technique to Train and Care for Your Favorite Animal.* New York: Viking, 1992.

Wise, Anna. "Biofeedback Meditation and the Awakened Mind". *The Art of Survival: A Guide to Yoga Therapy.* (M.L. Gharote and Maureen Lockhart, eds.) London: Unwin, 1987.

INDEX

ABOUT THE AUTHOR

ANNA WISE, diplomate in Neurotherapy, diplomate in Peak Performance, and certified by the Neurotherapy and Biofeedback Certification Board (CNBC), is one of the world's leading authorities on the use of real-time dynamic EEG for achieving high-performance and meditative states. Wise trained and worked in London, England, for eight years with the late C. Maxwell Cade, scientist and psycho-biologist. She holds an M.A. in humanistic psychology, and was a founding board member of the European Association for Humanistic Psychology, based in Geneva, Switzerland.

Anna Wise is the author of *Awakening the Mind: A Guide to Mastering the Power of Your Brainwaves* (Tarcher/Penguin, 2002), *The High Performance Mind: Mastering Brainwaves for Insight, Healing, and Creativity* (Tarcher/Penguin, 1995), and contributed to *The Art of Survival: A Guide to Yoga Therapy* (Unwin Hyman, 1987) with her chapter, "Biofeedback Meditation and the Awakened Mind." Wise also has innovated the complex use of "frequency scores," combining these scores with music to encourage deep meditation and a high-performance mind. She has collaborated with Dr. Andrew Weil as the frequency composer on their CD, *Sound Body, Sound Mind: Music for Healing,* as well as producing other music CDs for meditation as well.

Anna Wise writes, lectures, and consults. She has taught biofeedback meditation and brainwave training for the past three decades and she has led workshops and seminars throughout the United States, including at Esalen Institute in Big Sur, California, as well as in Europe, South America, and Asia. She can be contacted at www.annawise.com.